The
FORESTER'S
MANUAL

The Forest Trees Of Eastern North America

ERNEST THOMPSON SETON

British Library Cataloguing-in-Publication Data
A catalogue record for this book is available from the
British Library

Ernest Thompson Seton

Ernest Thompson Seton was born on 14th August 1860, in South Shields, County Durham, England. He grew up to be a pioneering author, wildlife artist, founder of the Woodcraft Indians, and one of the originators of the Boy Scouts of America (BSA).

The Seton family emigrated to Canada when Ernest was just six years old, and most of his childhood was consequently spent in Toronto. As a youth, he retreated to the woods to draw and study animals as a way of avoiding his abusive father – a practice which shaped the rest of his adult life. On his twenty-first birthday, Seton's father presented him with a bill for all the expenses connected with his childhood and youth, including the fee charged by the doctor who delivered him. He paid the bill, but never spoke to his father again.

Originally known as Ernest Evan Thompson, Ernest changed his name to Ernest Thompson Seton, believing that Seton had been an important name in his paternal line. He became successful as a writer, artist and naturalist, and moved to New York City to further his career. Seton later lived at 'Wyndygoul', an estate that he built in Cos Cob, a section of Greenwich, Connecticut. After experiencing vandalism by some local youths, Seton invited the young miscreants to his estate for a weekend, where he told them what he claimed were stories of the American Indians and of nature.

After this experience, he formed the Woodcraft Indians (an American youth programme) in 1902 and invited the local youth to join (at first just boys, but later girls as well). The stories that Seton told became a series of articles written for the *Ladies Home Journal,* and were eventually collected in *The Birch Bark Roll of the Woodcraft Indians* in 1906. Seton also met Scouting's founder, Lord Baden-Powell, in 1906. Baden-Powell had read Seton's book of stories, and was greatly intrigued by it. After the pair had met and shared ideas, Baden-Powell went on to found the Scouting movement worldwide, and Seton became vital in the foundation of the Boy Scouts of America (BSA) and was its first Chief Scout (from 1910 – 1915). Despite this large achievement, Seton quickly became embroiled in disputes with the BSA's other founders, Daniel Carter Beard and James E. West.

In addition to disputes about the content of Seton's contributions to the Boy Scout Handbook, conflicts also arose about the suffrage activities of his wife, Grace, and his British citizenship (it being *an American* organization). In his personal life, Seton was married twice. The first time was to Grace Gallatin in 1896, with whom he had one daughter, Ann (who later changed her name to Anya), and secondly to Julia M. Buttree, with whom he adopted an infant daughter, Beulah (who also changed her first name, to Dee). Alongside his work with the Woodcraft Indians and the BSA, Seton also found time to pursue his primary interest – that of nature writing.

Seton was an early pioneer of animal fiction writing, his most popular work being *Wild Animals I Have Known* (1898), which contains the story of his killing of the wolf Lobo. He later became involved in a literary debate known as the nature fakers controversy, after John Burroughs published an article in 1903 in the *Atlantic Monthly* attacking writers of sentimental animal stories. The controversy lasted for four years and included important

American environmental and political figures of the day, including President Theodore Roosevelt. Seton was also associated with the Santa Fe arts and literary community during the mid-1930s and early 1940s, which comprised a group of artists and authors including author and artist Alfred Morang, sculptor and potter Clem Hull, painter Georgia O'Keeffe, painter Randall Davey, painter Raymond Jonson, leader of the Transcendental Painters Group, and artist Eliseo Rodriguez.

In 1931, Seton became a United States citizen. He died on 23rd October, 1946 (aged eighty-six) in Seton Village in northern New Mexico. Seton was cremated in Albuquerque. In 1960, in honour of his 100th birthday and the 350th anniversary of Santa Fe, his daughter Dee and his grandson, Seton Cottier (son of Anya), in a fitting tribute to the man who loved his surrounding countryside so much, scattered his ashes over Seton Village from an airplane.

THE FORESTERS' MANUAL

THE FORESTER'S MANUAL

Or

The Forest Trees

Of

Eastern North America.

No. 2 of Scout Manual Series

By

ERNEST THOMPSON SETON

CHIEF SCOUT

BOY SCOUTS OF AMERICA

GARDEN CITY   NEW YORK
DOUBLEDAY, PAGE & COMPANY
1912

PREFACE

THIS book is meant to be a Foresters' Manual, not a Botany. In it I aim to give the things that appealed to me as a boy: First the identification of the tree, second where it is found, third its properties and uses, and last, various interesting facts about it.

I have included much information about native dyes, because it is all in the line of creating interest in the trees; and because it would greatly improve our color sense if we could return to vegetable dyes, and abandon the anilines that have in many cases displaced them. So also because of the interest evoked as well as for practical reasons I have given sundry medical items; some of these are from H. Howard's "Botanic Medicine," 1850. Several of the general notes are from George B. Emerson's "Trees and Shrubs of Massachusetts," 1846.

As starting point I have used Britton and Brown's "Illustrated Flora" (Scribner, 1896) and have got much help from Harriet L. Keeler's "Our Native Trees" (Scribner, 1900).

The illustrations were made by myself from fresh specimens in the woods, or in some cases from preserved specimens in the Museum of the New York Botanical Garden at Bronx Park.

The maps were made for this work by Mr. Norman Taylor, Curator of Plants in the Brooklyn Botanic Garden, N. Y., with corrections in Canada by Prof. John Macoun of the Geological Survey at Ottawa, Canada.

To Dr. N. L. Britton, Mr. Norman Taylor, and Prof. John Macoun, I extend my hearty thanks for their kind and able assistance.

The names of trees are those used in Britton's "North American Trees," 1908.

CONTENTS

vii

CONTENTS

INTRODUCTION

ALL the common forest trees of the region defined are given herein. I have, however, omitted a few rare stragglers on the South and West and certain trees that are big in the Gulf States but mere shrubs with us.

Remember when using this list as a key, that you will not often find a leaf exactly like the one in the book; look rather for an illustration of the same *general character* as the one in your hand; place your leaf with the one *most nearly* like it. Avoid the leaves of stump-sprouts and saplings; they are rarely typical; and especially get the fruit when possible; "*the tree is known by its fruit.*" In some cases nothing but the fruit can settle what your species is.

In each (with five exceptions) the fruit is given of exact natural size. The exceptions are the Osage Orange or Bodarc, the Mountain Magnolia, Red-bud, Honey Locust, and Kentucky Coffee-tree, all of which are given in half size.

In giving the weight of each kind of timber it is assumed to be dry and seasoned. All of our woods are lighter than water when seasoned; but many of them sink when green. The heaviest of our list is Yellow Oak, 54 ℔s. per cubic foot; the lightest is Northern Cedar, 20 ℔s. A cubic foot of water weighs 63 ℔s., and for further interesting comparison, a cubic foot of iron weighs 470 ℔s., lead 718 ℔s., gold 1228 ℔s., and platinum, 1323 ℔s.

ONE HUNDRED OF THE BEST KNOWN NATIVE TIMBER TREES OF NORTHEASTERN AMERICA

(That is, North America east of Long. 100° west, and north of North Lat. 36°)

1. PINACEÆ — CONIFERS OR PINE FAMILY

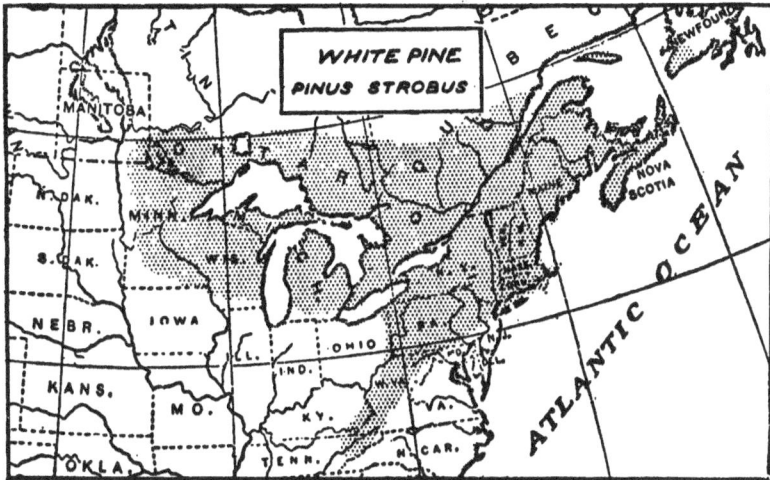

WHITE PINE
PINUS STROBUS

WHITE PINE, WEYMOUTH PINE. (*Pinus Strobus*)

A noble evergreen tree, up to 175 feet high. The lumberman's prize. Its leaves are in bunches of 5, and are 3 to 5 inches long; cones 4 to 8 inches long. Wood pale, soft, straight-grained, easily split. Warps and checks less than any other of our timbers. A cubic foot weighs 24 lbs.

Pine knots are hard masses of rosin, they practically never rot; long after the parent log is reduced to dust by the weather, the knots continue

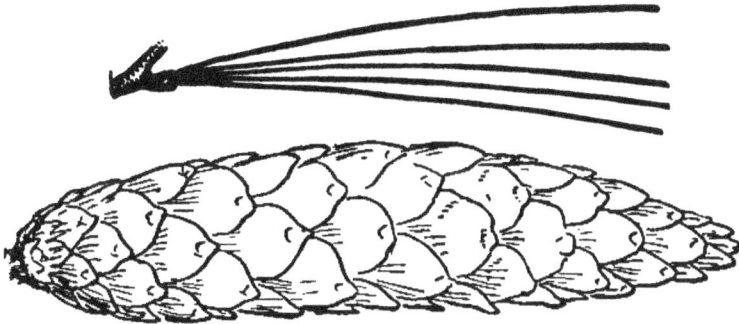

hard and sound. They burn freely with hot flame and much smoke
and are the certain fuel for a fire in all weathers. In a less degree the
same remarks apply to the larger roots.

RED PINE, CANADIAN PINE, NORWAY PINE. (*Pinus resinosa*)

Evergreen; somewhat less than the White Pine, with leaves 4 to 6
inches long, in bunches of 2, cones 1½ to 2½ inches long. Wood
darker, harder and heavier. A cubic foot weighs 30 lbs.

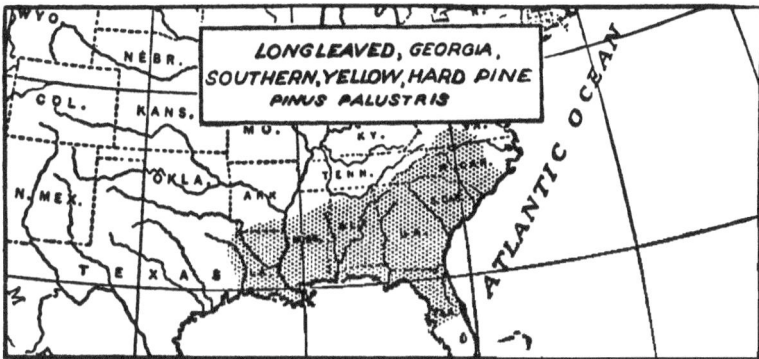

LONG-LEAVED PINE, GEORGIA PINE, SOUTHERN PINE, YELLOW PINE, HARD PINE. (*Pinus palustris*)

A fine tree, up to 100 feet high; evergreen; found in great forests in the Southern States; it supplies much of our lumber now; and most of our turpentine, tar and rosin. Wood strong and hard, a cubic foot weighs 44 lbs. Its leaves are 10 to 16 inches long, and are in bunches of 3's; cones, 6 to 10 inches long.

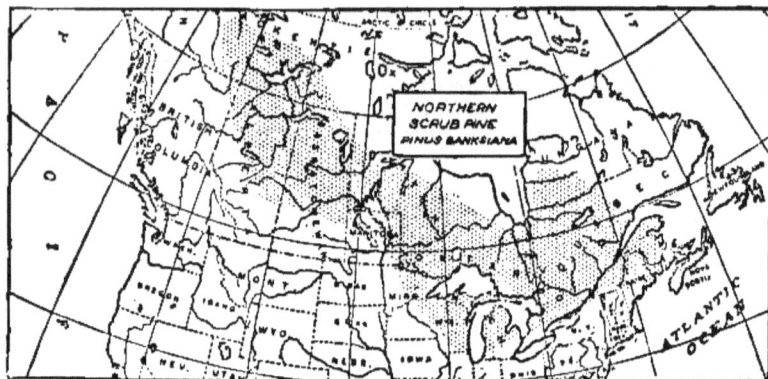

JACK-PINE, BANKSIAN PINE, GRAY PINE, LABRADOR PINE, HUDSON BAY PINE, NORTHERN SCRUB PINE. (*Pinus Banksiana*)

Evergreen; 40 to 60 feet high; rarely 100. Leaves in bunches of 2, and 1 to 2½ inches long; cone, 1 to 2 inches long. Dr. Robt. Bell of Ottawa says its seeds germinate better when the cone has been scorched. Wood, soft, weak. A cubic foot weighs 27 lbs.

In 1907 on Great Slave River, N. latitude 60, we cut down a Jack-pine

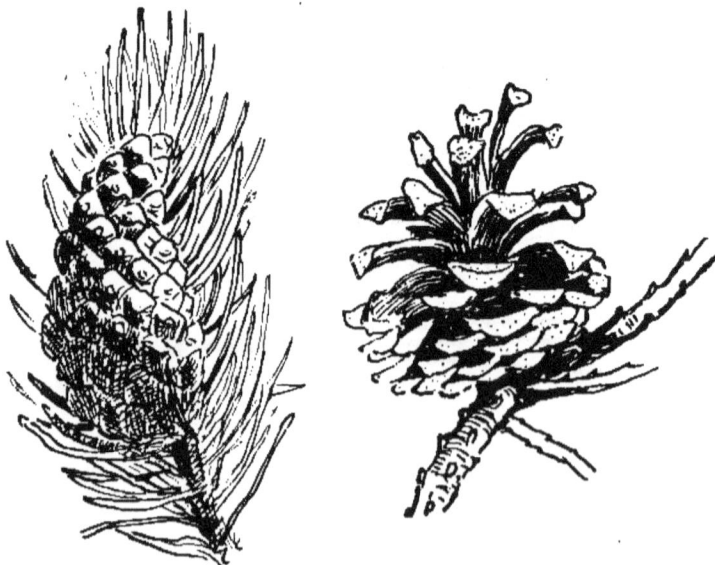

12 feet high, it was one inch thick and had 23 rings at the bottom. Six feet up it had 12 rings and 20 whorls — in all it appeared to have 43 whorls, of these 20 were on the lower part. This tree grew up in a dense thicket under great difficulties and was of very slow growth, the disagreement between rings and whorls was puzzling.

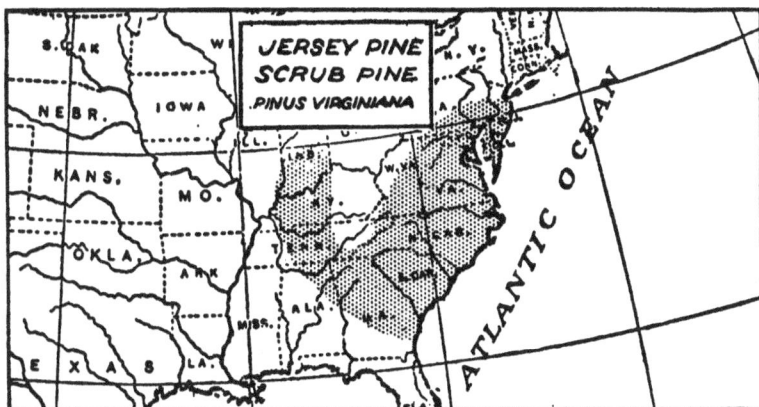

JERSEY PINE, SCRUB PINE. (*Pinus virginiana*)

Usually a small tree. Leaves 1½ to 2 inches long and in bunches of 2's; cones 1½ to 2½ inches long. Wood soft, weak, light orange; a cubic foot weighs 33 lbs. In sandy soil.

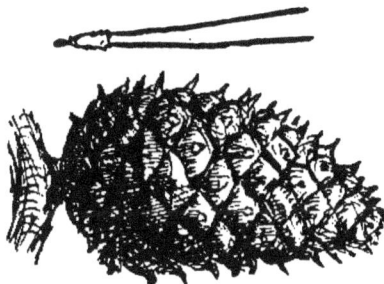

YELLOW SPRUCE,
SHORT-LEAVED, BULL PINE
PINUS ECHINATA

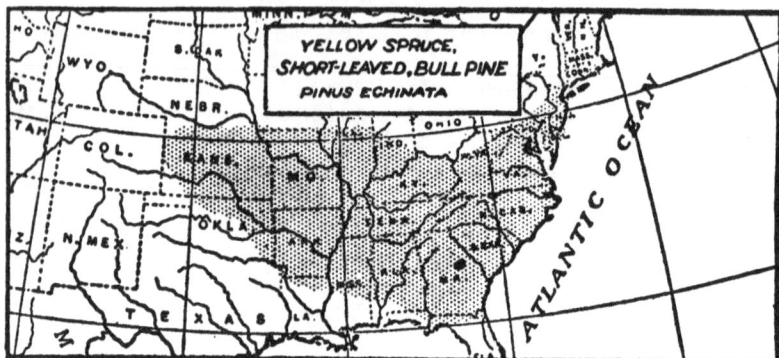

Y‌ELLOW P‌INE, S‌PRUCE P‌INE, S‌HORT-L‌EAVED P‌INE, B‌ULL P‌INE.   (*Pinus echinata*)

A forest tree, up to 100 feet high.   Leaves 3 to 5 inches long, and in bunches of 2's or 3's; cones about 2 inches long.   Wood heavy, strong, orange; a cubic foot weighs 38 lbs.   Valuable timber.

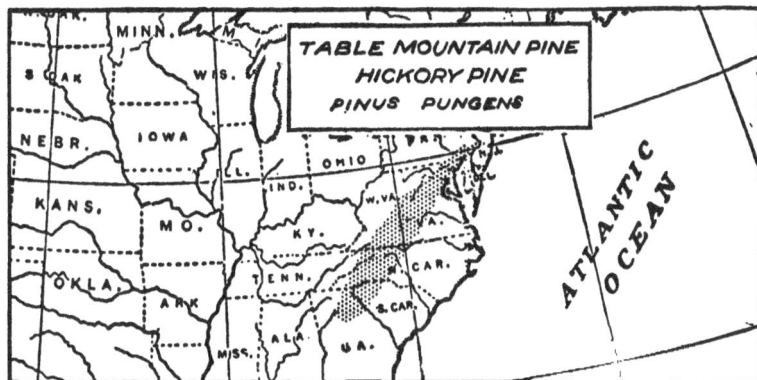

TABLE MOUNTAIN PINE
HICKORY PINE
PINUS PUNGENS

TABLE MOUNTAIN PINE, HICKORY PINE, (*Pinus pungens*)

A small tree, rarely 60 feet; leaves 2½ inches long; mostly in bunches of 2's or sometimes 3's; cones 3½ to 5 inches long. In the mountains New Jersey to North Carolina. Wood, weak, soft, brittle, a cubic foot weighs 31 lbs.

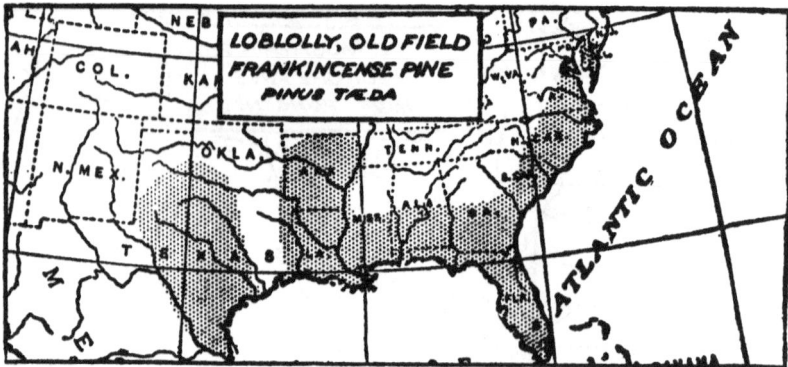

LOBLOLLY, OLD FIELD PINE, FRANKINCENSE PINE. (*Pinus Tæda*)

A fine forest tree, up to 150 feet. Leaves 6 to 10 inches long, and in bunches of 3's, rarely 2's; cones 3 to 5 inches long. Wood, weak, brittle, coarse, light brown, a cubic foot weighs 34 lbs.

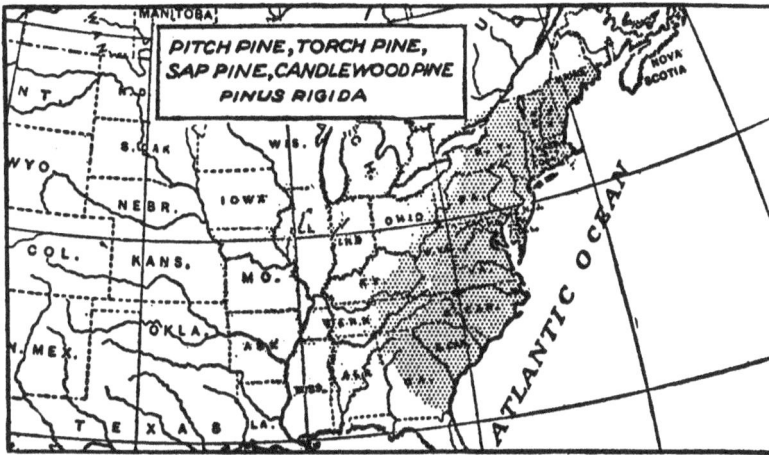

PITCH PINE, TORCH PINE, SAP PINE, CANDLEWOOD PINE.   (*Pinus rigida*)

A small tree, rarely 75 feet high; evergreen; leaves 3 to 5 inches long and in clusters of 3, rarely 4; cones 1½ to 3 inches long.   So charged with resin as to make a good torch.   Remarkable for producing shoots from stumps.   Wood, soft, brittle, coarse-grained, and light.   A cubic foot weighs 32 lbs.   "It is the only pine that can send forth shoots after injury by fire."   (*Keeler*).   The pine of the "pine-barrens" of Long Island and New Jersey.

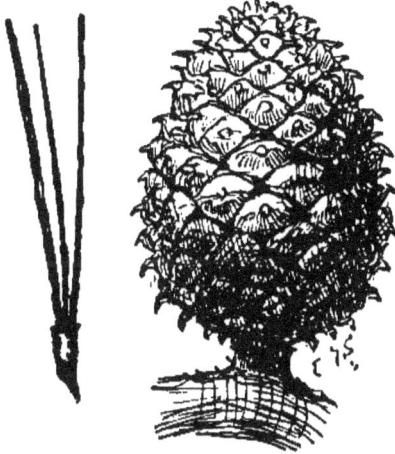

TAMARACK, LARCH OR HACKMATACK. (*Larix laricina*)

A tall, straight, tree of the northern swamps yet often found flourishing on dry hillsides. One of the few conifers that shed all their leaves each fall. Leaves ½ to 1 inch long; cones ½ to ¾ inch. Wood very resinous heavy and hard, "a hard, soft wood" very durable as posts, in Manitoba I have seen tamarack fence posts unchanged after twenty years' wear. It is excellent for firewood, and makes good sticks for a rubbing stick fire. A cubic foot weighs 39 lbs. Found north nearly to the limit of trees; south to northern New Jersey and Minnesota.

WHITE SPRUCE. (*Picea canadensis*)

Evergreen; 60 to 70 or even 150 feet high. Leaves ½ to ¾ inch long; cones 1½ to 2 inches long, are at the tips of the branches and deciduous; the twigs smooth. Wood white, light, soft, weak, straight-grained, not durable; a cubic foot weighs 25 lbs. Its roots afford the *wattap* or cordage for canoe-building and camp use generally.

Spruce roots to be used as "*wattap*" for lacing a canoe, making birch-bark vessels or woven baskets, may be dug up at any time and kept till needed.

An hour before using, soak in hot water till quite soft. They should be cleared of the bark and scrubbed smooth. *Beautiful and strong baskets* may be made of this material. It may be colored by soaking in dyes made as follows:

Red by squeezing the juice out of berries, especially *blitum* or squaw-berries.

Dull red by soaking in strong tea made from the pink middle bark of hemlock.

Black can be boiled out of smooth red sumac or out of butternut bark.

Yellow by boiling the inner bark of black oak or the root of gold seal or hydrastis.

Orange by boiling the inner bark of alder, of sassafras or of the yellow oak.

Scarlet by first dyeing yellow, then dipping in red.

Nearly every tree bark, root bark and fruit has a peculiar dye of its own which may be brought out by boiling, and intensified with vinegar, salt, alum, iron or uric salts. Experiments usually produce surprises.

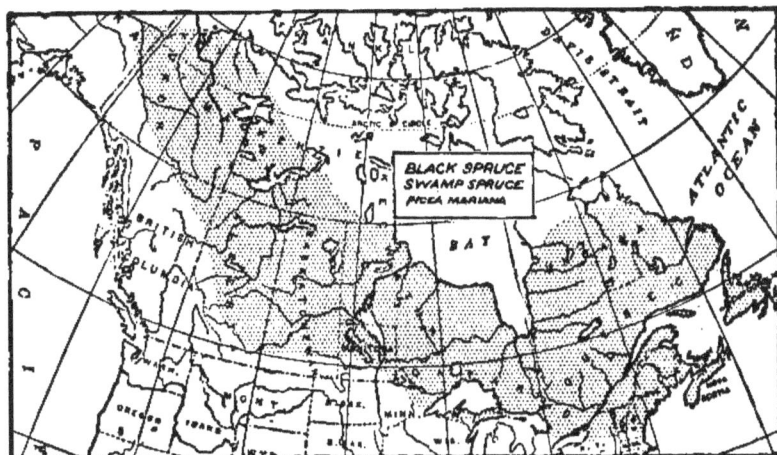

BLACK SPRUCE. (*Picea Mariana*)

Evergreen. Somewhat smaller than the preceding, rarely 90 feet high, with small rounded cones 1 to 1¼ inches long; they are found near the trunk and do not fall off; edges of scales more or less indented. In their September freshness the cones of Black Spruce are like small purple plums and those of White Spruce like small red bananas; twigs, stout and downy; wood and roots similar to those of White Spruce. Leaves about ½ inch long with rounded tops.

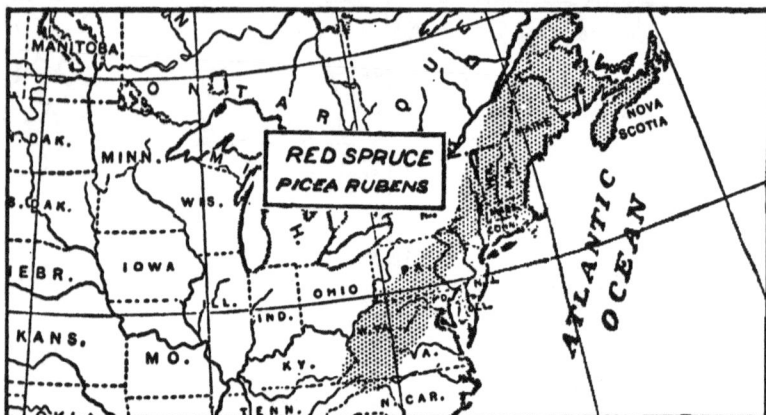

RED SPRUCE.   (*Picea rubens*)

Evergreen. Much like the Black Spruce but with larger, longer cones about 1½ inch long and red when young, they are half way between tip and trunk on the twigs; edges of scales smooth and unbroken; twigs slender, leaves sharp pointed. Roots as in White Spruce, but wood redder and weigh 28 lbs. An eastern tree. In many ways half way between the White and Black Spruces.

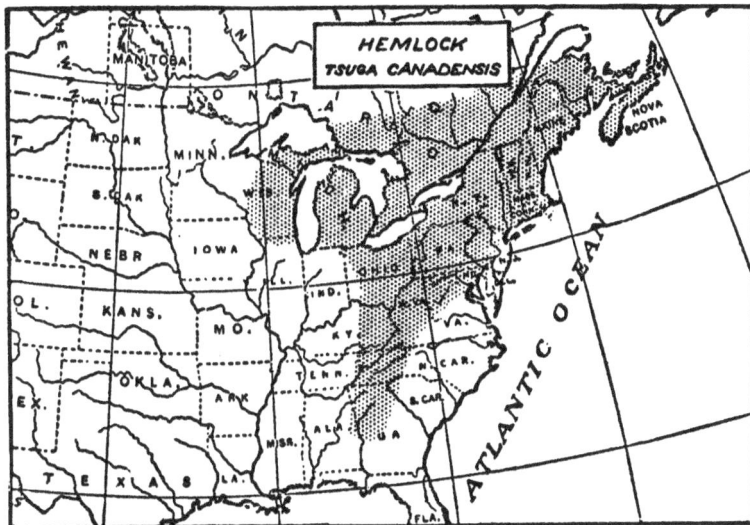

HEMLOCK.   (*Tsuga canadensis*)

Evergreen; 60 to 70 feet high; occasionally 100; wood pale, soft, coarse, splintery, not durable. A cubic foot weighs 26 lbs. Bark full of tannin. Leaves ½ to ¾ inch long; cones about the same. Its knots are so hard that they quickly turn the edge of an axe or gap it as a stone might; these are probably the hardest vegetable growth in our woods. It is a tree of very slow growth — growing inches while the White Pine is putting forth feet. Its topmost twig usually points easterly. Its inner bark is a powerful astringent. A tea of the twigs and leaves is a famous woodman's sweater.

"As it bears pruning to almost any degree without suffering injury, it is well suited to form screeens for the protection of more tender trees and plants, or for concealing disagreeable objects.

"But the most important use to which this bark is applied, and for which it is imported from Maine, is as a substitute for oak bark in the preparation of leather. It contains a great quantity of tannin, combined with a coloring matter which gives a red color to the leather apt to be communicated to articles kept long in contact with it." (*Emerson.*)

There is another species in the South (*T. Caroliniana*) distinguishable by its much larger cones.

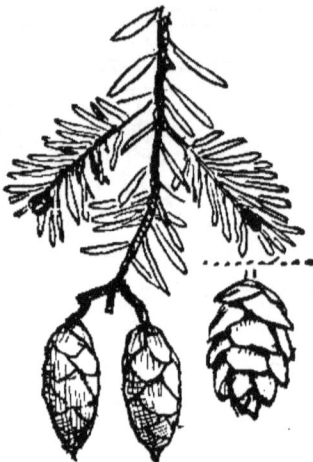

Twig and cones of Hemlock (life size)

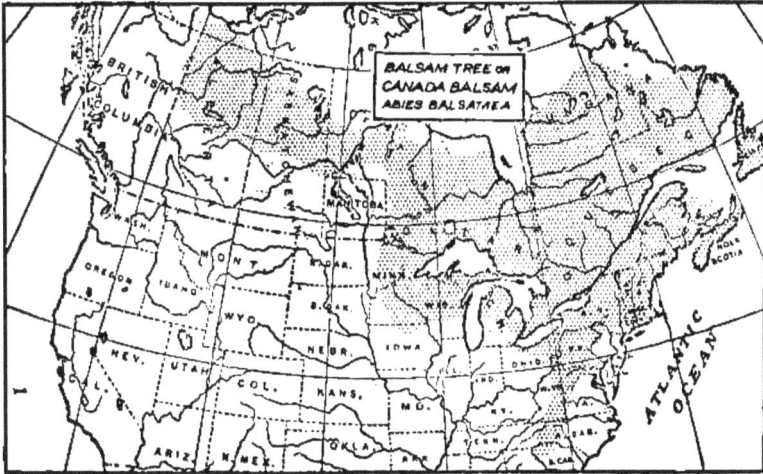

BALSAM TREE OR CANADA BALSAM.   (*Abies balsamea*)

Evergreen; famous for the blisters on its trunk, yielding Canada Balsam which makes a woodman's plaster for cuts or a waterproof cement;
and for the exquisite odor of its boughs, which also supply the woodmen's
ideal bed.   Its *flat* leafage is distinctive.   Wood pale, weak, soft,
perishable.   A cubic foot weighs 24 lbs.   The name "balsam" was given

because its gum was long considered a sovereign remedy for wounds, inside and out. It is still used as a healing salve. In the southern Alleghanies is a kindred species (*A. fraseri*) distinguished by silvery underside of leaves, and smaller rounder cones.

The Conifers illustrate better than others of our trees tne process and plan of growth. Thus a seedling pine has a tassel or two at the top of a slender shoot, next year it has a second shoot from the whorl that finished last year. So each year there is a shoot and a whorl corresponding exactly with its vigor that season, until the tree is so tall that the lower whorls die, and their knots are overlaid by fresh layers of timber. The timber grows smoothly over them, but they are there just the same, and any one carefully splitting open one of these old forest patriarchs, can count on the spinal column the years of its growth, and learn in a measure how it fared each season.

In working this out I once cut down and examined a tall Balsam in the Bitterroot Mountains of Idaho. It was 84 feet high, had 52 annual rings; and at 32 inches from the ground, that is, clear of the root bulge, it was 15 inches in diameter.

The most	growth	was	on	the	N.E.	side	of	the	stump	— 9 in.
"	next	"	"	"	"	E.	"	"	"	"	— 8½in.
"	"	"	"	"	"	S.	"	"	"	"	— 8 in.
"	"	"	"	"	"	N.	"	"	"	"	— 6½in.
"	"	"	"	"	"	W.	"	"	"	"	— 6½in.
"	least	"	"	"	"	N.W.	"	"	"	"	— 6 in.

There were 50 well-marked whorls and 20 not well marked; there were altogether 70 whorls, but 20 were secondary. The most vigorous growth on the tree trunk corresponded exactly with the thickest ring of wood on the stump. Thus annual ring No. 33 on the stump counting from the centre coincided with an annual shoot of more than 2 feet length, which would be that of the wet season of 1883. Some of the annual shoots were but 6 inches long and had correspondingly thin rings. There was, of course, one less ring above each whorl or joint.

Similar studies made on Jack Pine and Yellow Pine gave similar results.

On hardwood trees especially those of alternate foliage one cannot so study them except when very young.

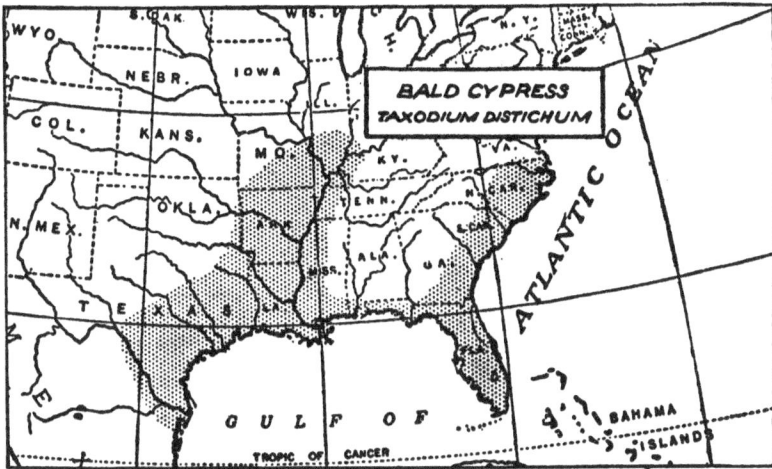

BALD CYPRESS.   (*Taxodium distichum*)

A fine forest tree, up to 150 feet, with thin leaves somewhat like those of Hemlock, half an inch to an inch long; cones rounded about an inch through.   Sheds its leaves each fall so is "bald" in winter, noted for the knees or upbent roots that it develops when growing in water. Timber soft, weak, but durable and valuable; a cubic foot weighs 27 lbs.   In low wet country.

ARBOR-VITÆ OR WHITE CEDAR. (*Thuja occidentalis*)

Evergreen, 50 or 60 feet high. Wood soft, brittle, coarse grained, extremely durable as posts; fragrant and very light (the lightest on our list). Makes good sticks for rubbing stick fire. A cubic foot weighs only 20 lbs. The scale-like leaves are about 6 or 8 to the inch; the cone half an inch long or less. There is a kindred species (*Chamaecyparis thyoides*) of more southern distribution. It has much smaller cones and leaves.

The Northern or White Cedar is noted for the dense thickets it forms in the hollows and hillsides of the eastern Canadian region. These banks, like evergreen hedges, are so close that they greatly modify the winter climate within their bounds — outside there may be a raging blizzard that no creature can face, while within all is dead calm and the frost less intense. The Cedar feeds its protegés too, for its evergreen boughs and abundant nuts are nutrient food despite their rosin smell and taste. Never do the deer and hares winter better than in cedar cover, and if there is great thicket in their region, they surely gather there as sparrows at a barn, or as rats around a brewery.

Enlarged leaves
Twigs and cones of Northern Arbor-vitæ

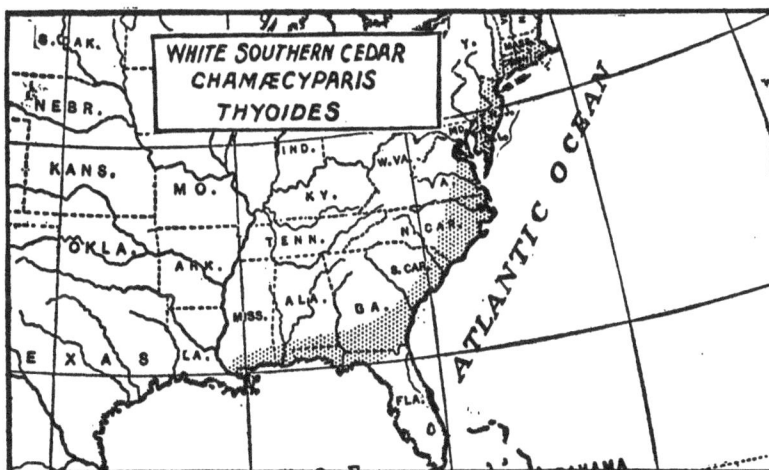

WHITE SOUTHERN CEDAR
CHAMÆCYPARIS
THYOIDES

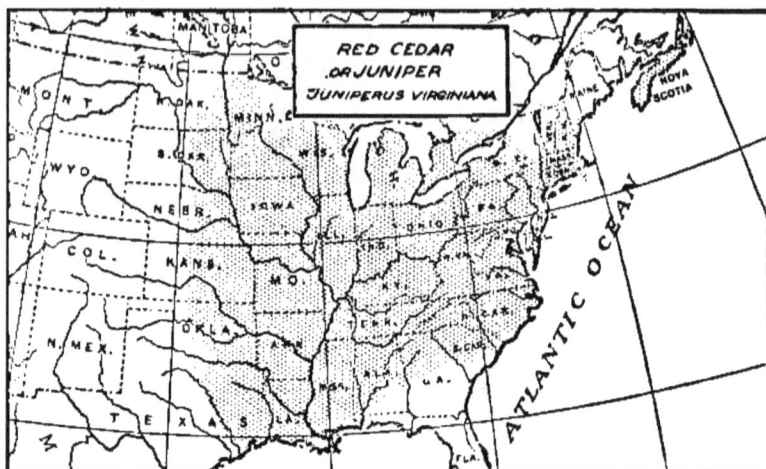

RED CEDAR OR JUNIPER. (*Juniperus Virginiana*)

Evergreen. Any height up to 100 feet. Wood, heart a beautiful bright red; sap wood nearly white; soft, weak, but extremely durable as posts, etc. Makes good sticks for rubbing stick fire. The tiny scale-like leaves are 3 to 6 to the inch; the berry-like cones are light blue and a quarter of an inch in diameter.

The berries of the European species are used for flavoring gin, which word is an abbreviation of Juniper.

"The medicinal properties of both are the same (Savin, of Europe) a decoction of the leaves having a stimulating effect, when used internally in cases of rheumatism and serving to continue the discharge from blisters, when used in the composition of cerate for that purpose." (*Emerson.*)

A cubic foot weighs 31 lbs.

Red Cedar showing fruit and two styles of twigs (life size)
on the same tree

2. SALICACEÆ—THE WILLOW FAMILY

The Willows are a large and difficult group. Britton and Brown
enumerate 34 species in the limits of northeastern America, and 160
on the globe, of which 80 are found in this continent. Of the 34, 9
only attain the dignity of trees. These are Ward's Willow, Peach-
leaved Willow, Shining Willow, Weeping Willow, Purple Willow, Mis-
souri Willow and the three herein described.

Of the shrubs, two only have a special interest in woodcraft, the Pussy-
Willow, because of its spring bloom, and the Fish-Net or Withy Willow.

Since the fruits of the Willows are born of catkins and are exceed-
ingly small and difficult of study, they are not figured.

BLACK WILLOW. (*Salix nigra*)

The common Willow of stream-banks, usually 20 to 40 feet high, sometimes 100. Bark nearly black. Its long, narrow, yellow-green shining leaves are sufficiently distinctive. A decoction of Willow bark and root is said to be the best known substitute for quinine. Noted for early leafing and late shedding; leaves 3 to 6 inches long. Wood pale, weak, soft, close-grained; a cubic foot weighs 28 lbs.

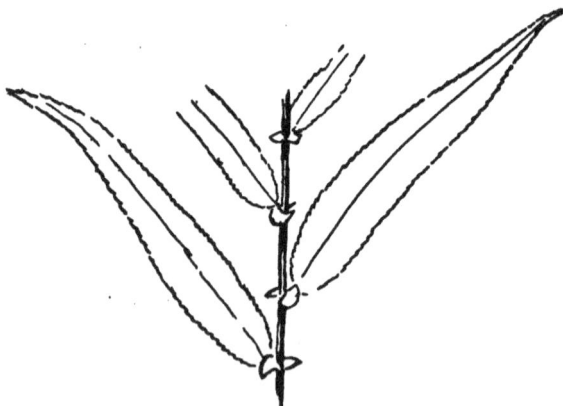

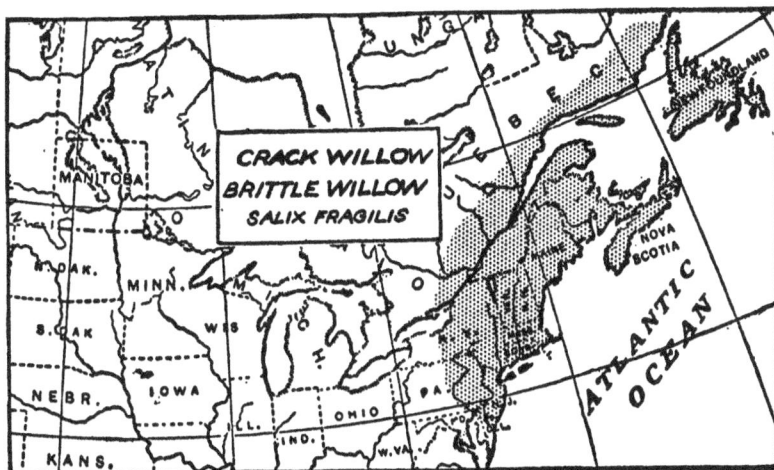

CRACK WILLOW, BRITTLE WILLOW. (*Salix fragilis*)

A tall slender tree, up to 80 feet high. Called "Crack" etc., because its branches are so much broken by the storms; too brittle for basket work, but a favorite for charcoal used in manufacture of gunpowder, etc. Its leaves, 4 to 7 inches long, are very distinctive. This is a European species but now thoroughly naturalized in the Northeastern States.

As a rough general rule the shape of the perfect tree is closely fashioned on that of the perfect leaf, for obviously they are the same material impelled by similar laws of growth, but we have two notable exceptions in the Lombardy Poplar and the common Willow. To conform to the rule these two leaves should change places.

GOLDEN WILLOW, GOLDEN OSIER, YELLOW WILLOW OR WHITE WILLOW
(*Salix alba*)

This is a tall tree, up to 90 feet high. Leaves 2 to 4½ inches long. It is the well known willow of dams; conspicuous in spring for the mass of golden rods it presents. It comes near being evergreen as it leafs so early and sheds so late, that it is bare of leaves for less than four months. Noted for its wonderful vitality and quickness of growth. Any living branch of it stuck in the ground soon becomes a tree. On the dam at Wyndygoul are large Willows, one of them 61 inches in circumference a foot from the ground though they were mere switches when planted eight years ago. A native of Europe, now widely naturalized in the Northeastern States and southern Canada.

PUSSY WILLOW OR GLAUCOUS WILLOW. (*Salix discolor*)

Usually a shrub, occasionally a tree, up to 25 feet high. Noted for its soft round catkins an inch long and two thirds of an inch thick, that appear in early spring before the leaves. The name Pussy is given either on account of these Catkins (little cats) or from the French "Poussé" budded.

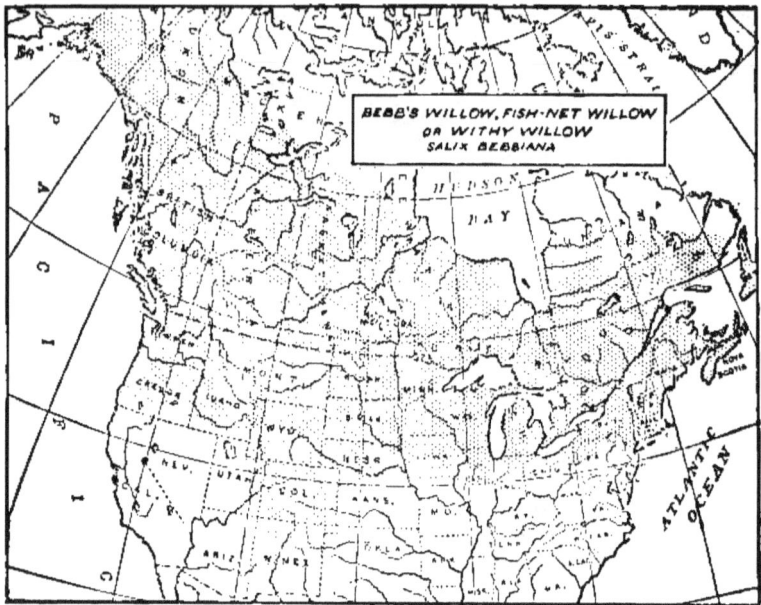

FISH-NET WILLOW OR WITHY WILLOW, BEBB'S WILLOW. (*Salix Bebbiana*)

This is a low thick bush or rarely a tree 20 feet high. It abounds near water, which seems a natural fitness, for its inner bark supplies the best native material for fish lines and fish nets in the North. It is called Withy Willow because its tough, pliant stems are used by farmers for withies or coarse cordage, especially for binding fence rails and stakes; though soft and pliant when put on they soon turn to horny hardness and last for years. Arctic to British Columbia north to Mackenzie River south to Pennsylvania and Utah.

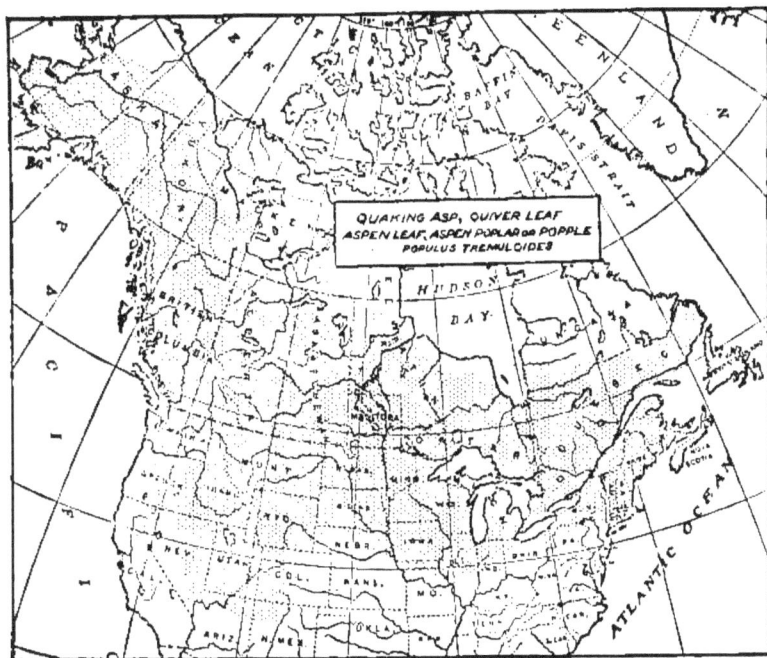

QUAKING ASP, QUIVER LEAF, ASPEN POPLAR OR POPPLE. (*Populus tremuloides*)

A small forest tree, but occasionally 100 feet high. Readily known by its smooth bark, of a light green or whitish color. The wood is pale, soft, close-grained, weak, perishable, and light. A cubic foot weighs 25 lbs. Good only for paper pulp, but burns well, when seasoned. When green it is so heavy and soggy that it lasts for days as a fire check or back-log. Leaves 1½ to 2 inches long. A tea of the bark is a good substitute for quinine, as tonic, cold cure, bowel cure and fever driver.

"Pieces of wood 2⅜ inches square, were buried to the depth of one inch in the ground, and decayed in the following order: Lime, American Birch, Alder and Aspen, in three years; Willow, Horse-Chestnut and Plane, in four years; Maple, Red Beech and Birch, in five years; Elm, Ash, Hornbeam and Lombardy Poplar in seven years; Robinia, Oak, Scotch Fir, Weymouth Pine, Silver Fir, were decayed to the depth of half an inch in seven years; while Larch, common Juniper, Virginia Juniper and Arbor-vitæ, were uninjured at the end of that time." *Balfour's Manual of Botany, 1855. P. 45.*

Quaking Asp

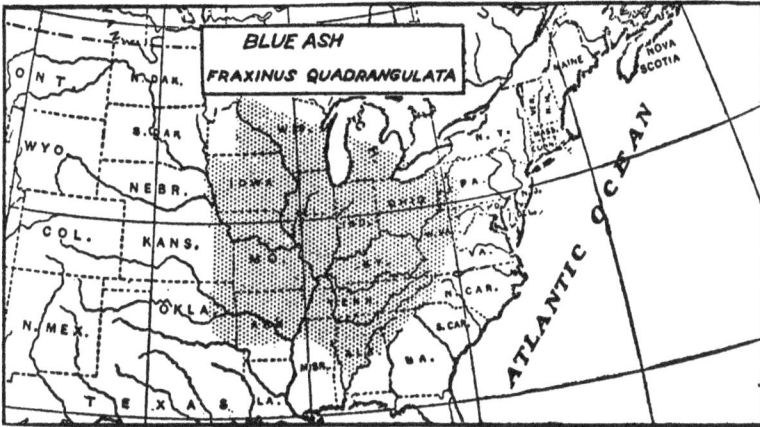

BLUE ASH.   (*Fraxinus quadrangulata*)

A tall tree of the Mississippi Valley, over 100 feet high.   Wood light yellow, hard, close, heavy.   A cubic foot weighs 45 lbs.   Leaflets 7 to 11, 3 to 5 inches long.   "The inner bark yields a blue color to water; hence its name."   "It may be distinguished among ashes by its peculiar, stout, four-angled, four-winged branches."   (*Keeler.*)

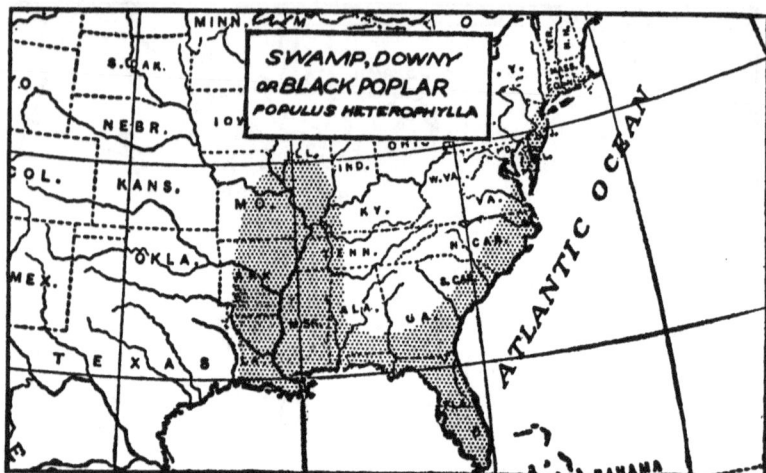

SWAMP, DOWNY OR BLACK POPLAR.   (*Populus heterophylla*)

A good-sized forest tree; up to 80 feet high.   A tree of cottonwood style; the young foliage excessively downy.   Wood soft, weak.   A cubic foot weighs 26 lbs.   Leaves 5 to 6 inches long.

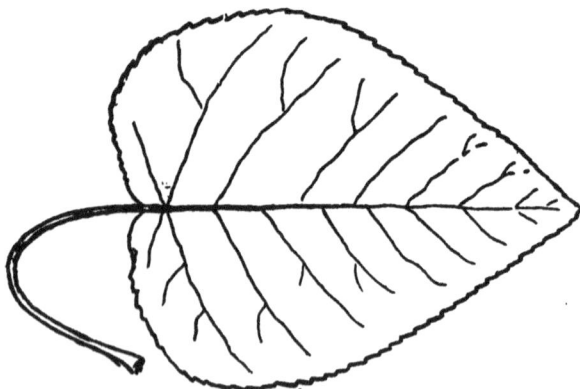

BALSAM POPLAR, BALM OF GILEAD, OR TACAMAHAC.   (*Populus balsami-
fera*)

Fifty or 60 feet ordinarily, but sometimes 100 feet high.   Bark rough
and furrowed.   The great size of the buds and their thick shiny coat
of fragrant gum are strong marks.   Wood much as in the preceding,
but weighs 23 lbs.   Leaves 3 to 6 inches long.   There is a narrow-
leafed form called *angustifolia*.

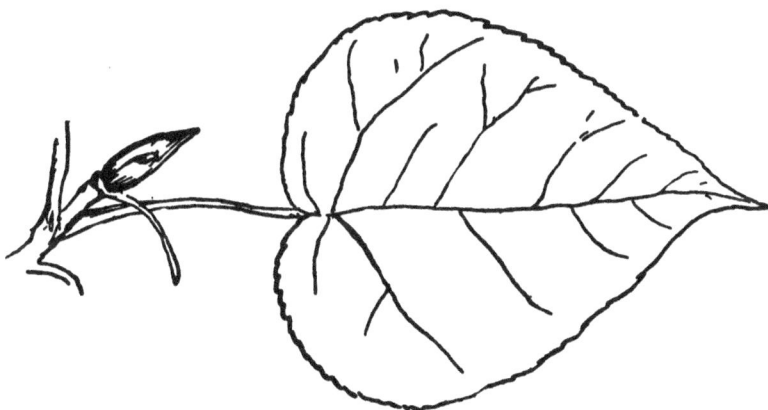

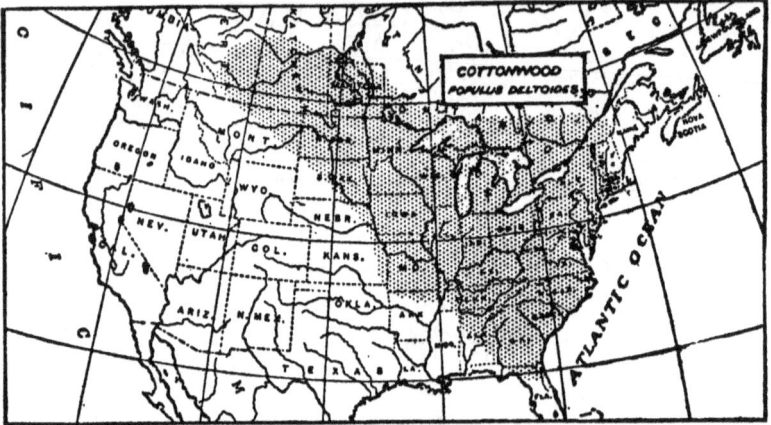

COTTONWOOD. (*Populus deltoides*)

Small and rare in the northeast. Abundant and large in west; even 150 feet high. Wood as in other poplars but weighs 24 lbs. Leaves 3 to 5 inches long. These and most of the poplars have the leaf stalks flattened laterally so that the slightest puff of wind vibrates the leaf, this with its shiny surface clears it of dust and enables it to live in dry places where different leaves would be stifled.

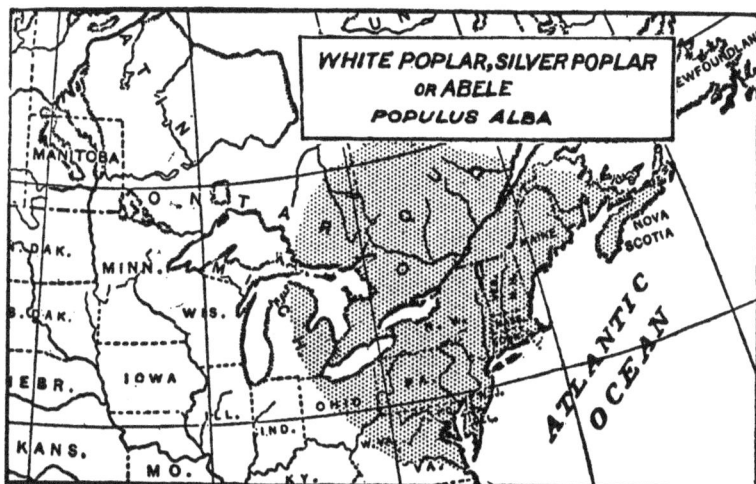

WHITE POPLAR, SILVER POPLAR OR ABELE. (*Populus alba*)

This is a species introduced from Europe. It is a tall forest tree; up to 120 feet. The dark glossy surface of the upper and the dense white velvet of the under side of leaves are strong features. Its wood is soft white and weighs 38 lbs. per cubic foot. Leaves 2½ to 4 inches long. Generally distributed in Northeastern States.

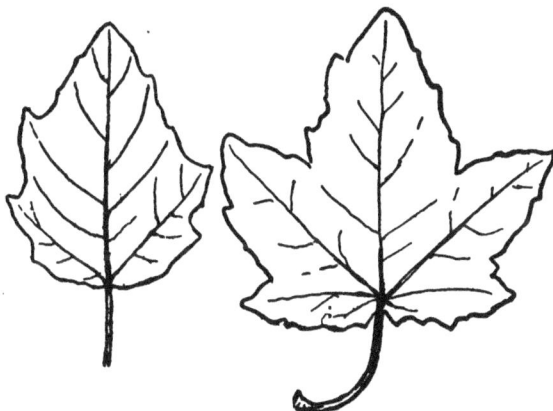

LOMBARDY POPLAR. (*Populus dilatata*)

Introduced from Europe. Its tall form is a familiar feature of the civilized landscape in Eastern America.

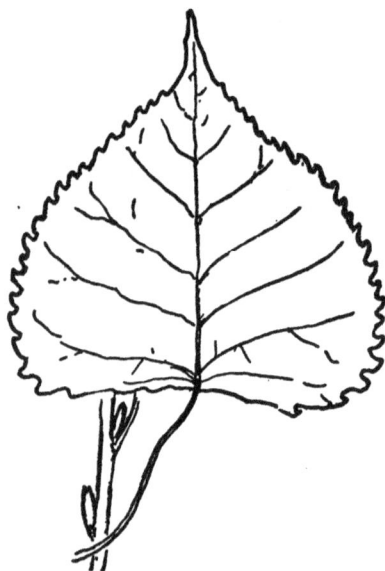

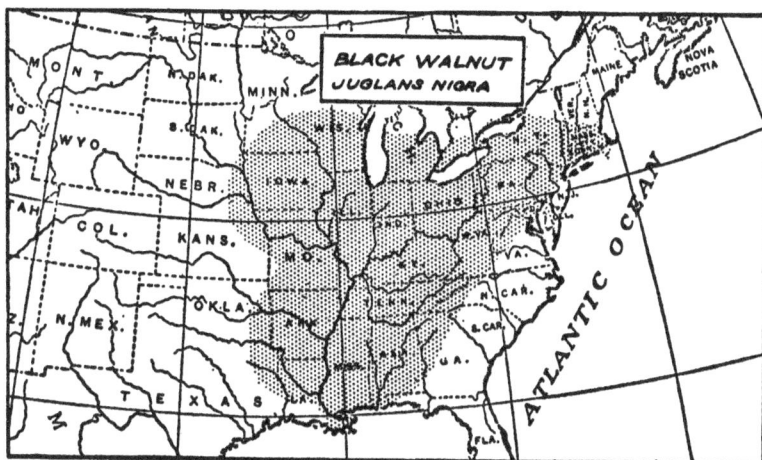

3. JUGLANDACEÆ OR WALNUT FAMILY

BLACK WALNUT.   (*Juglans nigra*)

A magnificent forest tree up to 150 feet high, usually much smaller in the east.   Wood, a dark purplish brown or gray; hard, close-grained; strong; very desirable in weather or ground work, and heavy.   A cubic foot weighs 38 lbs.   Leaflets 13 to 23; and 3 to 5 inches long.   Fruit nearly round, 1½ to 3 inches in diameter.

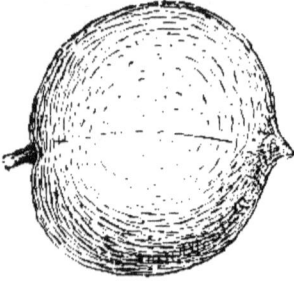

Fruit of Black Walnut   Fruit of Butternut

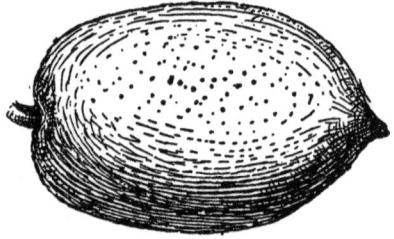

Both life size

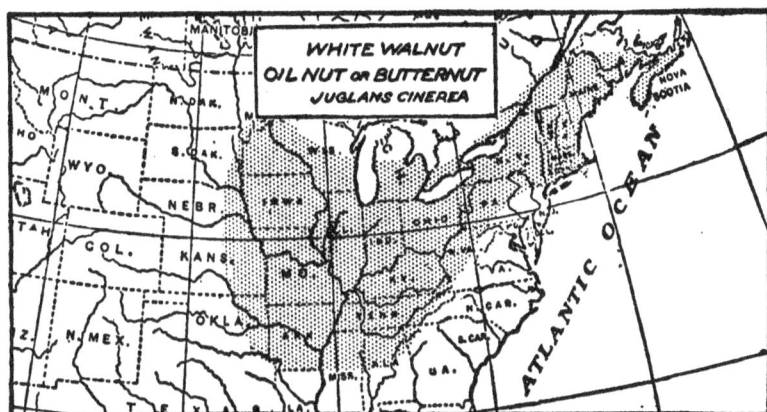

WHITE WALNUT, OIL NUT OR BUTTERNUT. (*Juglans cinerea*)

Much smaller than the last, rarely 100 feet high; with much smoother bark and larger, coarser, compound leaves, of fewer leaflets but the petioles or leaflet stalks and the new twigs are covered with sticky down.

"The bark and the nut are also used to give a brown color to wool. The Shakers at Lebanon dye a rich purple with it. Bancroft says that the husks of the shells of the Butternut and Black Walnut, may be employed in dyeing a fawn color, even without mordants. By means of them, however, greater brightness and durability are given to the color. The bark of the trunk gives a black, and that of the root a fawn color, but less powerful. From the sap an inferior sugar has been obtained. The leaves, which abound in acrid matter, have been used in the form of powder as a substitute for Spanish Flies." (*Emerson.*)

A decoction of the inner bark, preferably of the root, is a safe mild purge, a teaspoonful of it as dark as molasses is a dose.

The wood is light-brown, soft, coarse, not strong but very enduring in weather and ground work; light; leaves 15 to 30 inches long; leaflets 11 to 19 in number and 3 to 5 inches long; fruit oblong 2 to 3 inches long.

KEY TO THE HICKORIES OF NORTH AMERICA

SHAGBARKS

Bark hanging loose in broad plates; leaflets 5 to 7, broad; nut, ridged and sweet, (1) Common Shagbark.

Bark hanging loose in long narrow strips; leaflets 7 to 9; twigs, orange; foliage, downy; nut, much larger, (2) Big Shagbark.

Bark hanging loose in long narrow strips; leaflets 5 to 7; much like No. 1, but nuts not ridged, (3) Small fruited Shagbark.

RIDGED OR NET BARKS

Leaflets 11 to 15, very broad; nut smooth and without angles (4) Pecan.

Leaflets 7 to 9, very narrow, willow-like; nut smooth and without angles, (5) Bitternut.

Leaflets 9 to 13, very narrow, willow-like, top one very thin; nut with angles, (6) Water Hickory.

Leaflets 7 to 9, broad terminal bud ½ to ¾ inches long; nut with angles, (7) Mockernut.

Leaflets 3 to 7, very broad terminal bud ¼ to ½ inch long; nut with little or no angles, (8) Pignut.

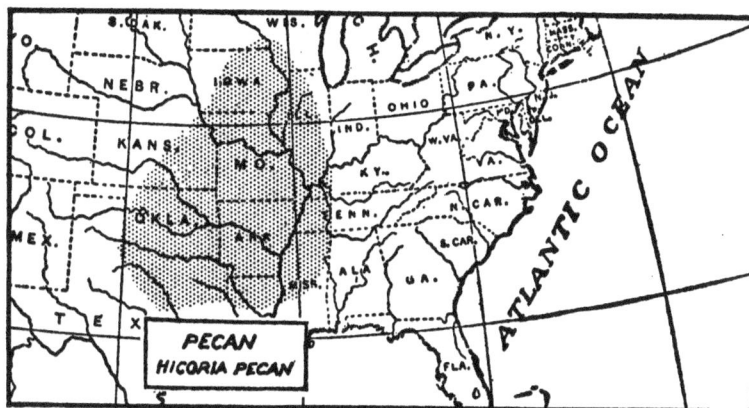

PECAN.   (*Hicoria Pecan*)

A tall slender forest tree in low moist soil along streams, up to 170 feet in height: famous for its delicious nuts, they are smooth and thin shelled; fruit, oblong, cylindrical, 1½ to 2½ inches long.   Its leaves are smooth when mature: leaflets 11 to 15, and 4 to 7 inches long: Wood hard and brittle, a cubic foot weighs 45 lbs.

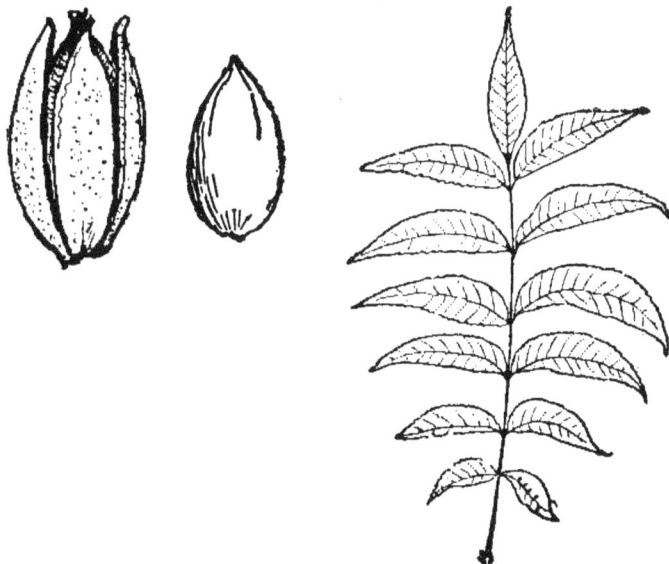

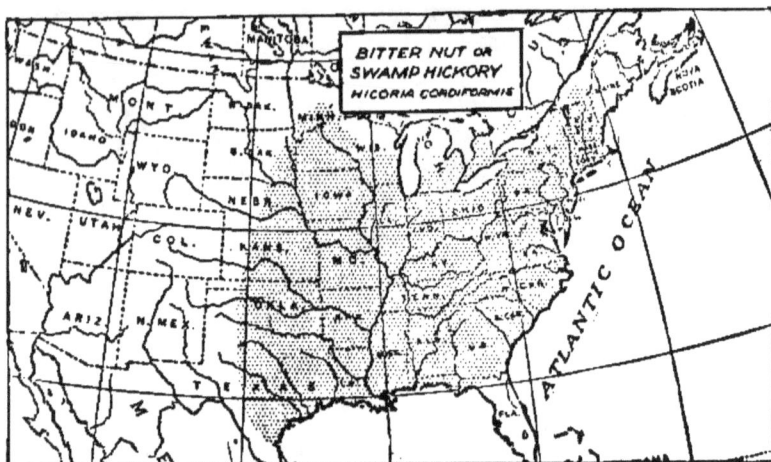

BITTERNUT OR SWAMP HICKORY. (*Hicoria cordiformis*)

A tall slender forest tree of low woods, up to 100 feet high; chiefly in Mississippi valley. Known by its small willow-like leaves; (7 to 9 leaflets); its close rough bark; its ridged fruit, and bitter kernel. Its leaves are 6 to 10 inches long, its leaflets 2 to 4 inches long. Wood, brownish, very hard, close-grained, tough, strong, and heavy; a cubic foot weighs 47 lbs. Excellent firewood.

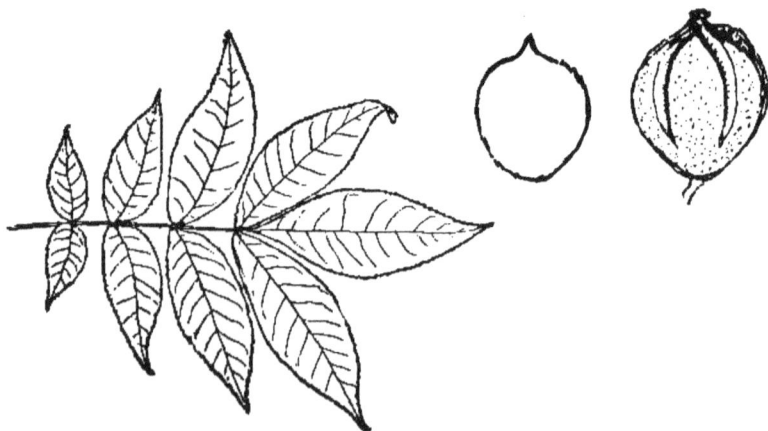

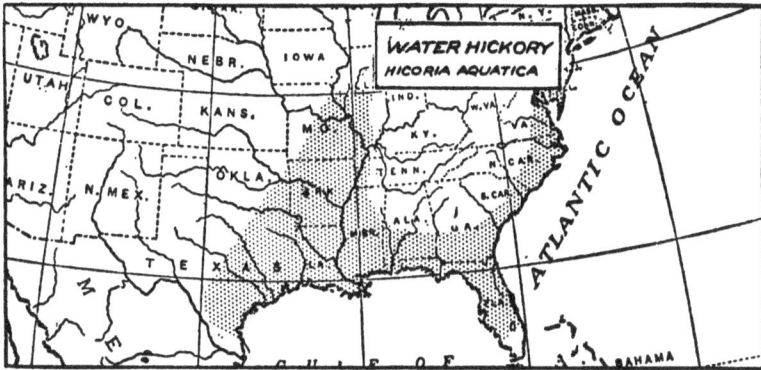

WATER HICKORY. (*Hicoria aquatica*)

A tall tree of southern swamps, up to 100 feet high; leaflets 9 to 13, 3 to 5 inches long, lance shaped, or the terminal one oblong; much like the Bitternut, but fruit longer and leaflets more numerous. Wood, soft, a cubic foot weighs 46 lbs. Virginia to Illinois, south to Texas and Florida.

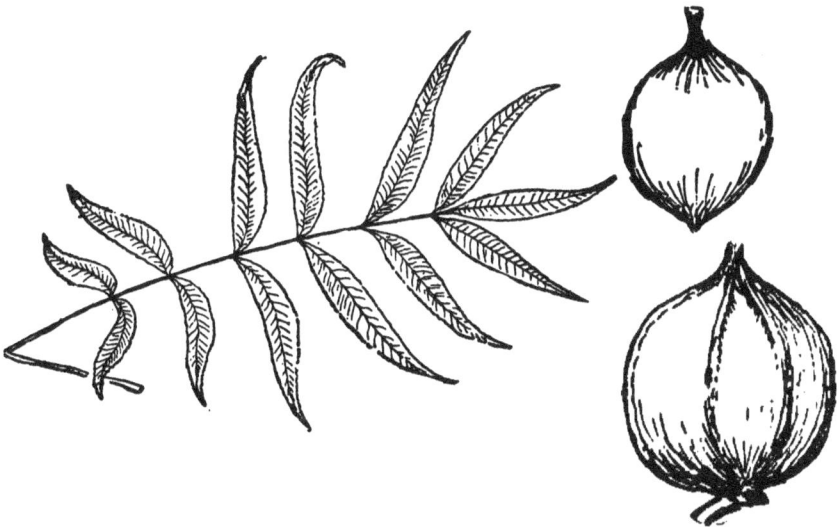

SHAGBARK, SHELLBARK
OR *WHITE HICKORY*
HICORIA OVATA

SHAGBARK, SHELLBARK OR WHITE HICKORY. (*Hicoria ovata*)

A tall forest tree up to 120 feet high. Known at once by the great angular slabs of bark hanging partly detached from its main trunk, forced off by the growth of wood, but too tough to fall. Its leaves are 8 to 14 inches long, with 5 to 7 broad leaflets. The wood is very light in color, close-grained, tough and elastic. It makes an excellent bow; is the best of fuel. A cubic foot weighs 52 lbs., so that it is the heaviest of the woods in this list except Post Oak, which is the same weight, and Yellow Oak, which is 2 lbs. heavier. It is the favorite for fork-handles and articles requiring strength and spring, but is useless for weather or ground work. Its nuts are the choicest of their kind. It is a tree of many excellencies.

THE BIG SHELL-BARK OR KING-NUT. (*Hicoria laciniosa*)

Ranges from Central New York south and westerly. It is much like the Shagbark but known by its downy young foliage and orange twigs; its leaflets 7 to 9, rarely 5 and very large, fruit 2 to 3 inches long and oblong, while in Shagbark they are 1¼ to 2¼ inches long and rounded. Wood 50 lbs. to cubic foot. In rich soil.

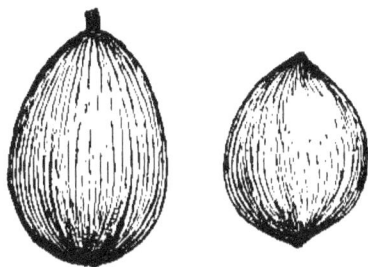

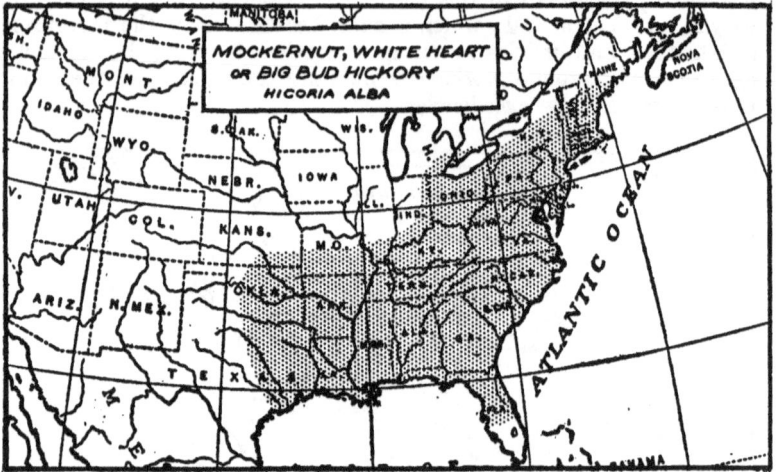

MOCKERNUT, WHITE HEART OR BIG-BUD HICKORY. (*Hicoria alba*)

A tall forest tree, up to 100 feet. Wood much like that of Shagbark, but not quite so heavy (51 lbs.). Its bark is smooth and furrowed like that of the Pignut. Its leaves like those of the Shagbark, but it has 7 to 9 leaflets, instead of 5 to 7; it has a large terminal bud ½ to ¾ of an inch long, and the leaves have a resinous smell. Its nut in the husk is nearly 2 inches long; the nut shell is 4-ridged toward the point, has a very thick shell and small sweet kernel.

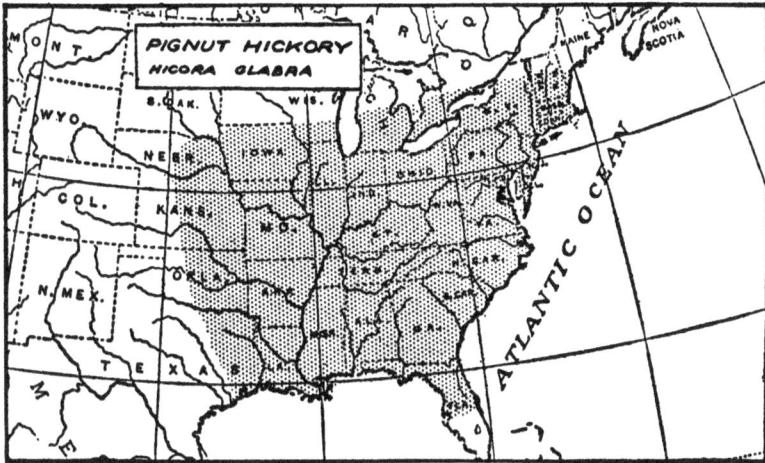

PIGNUT HICKORY. (*Hicoria glabra*)

A tall forest tree; 100 and up to 120 feet high. Wood much as in the Mockernut; bark smooth and furrowed; not loose plates. Leaves 8 to 12 inches long. Nut slightly or not at all angular, very thick shelled; the pear shape of fruit is a strong feature, 1¼ to 2 inches long.

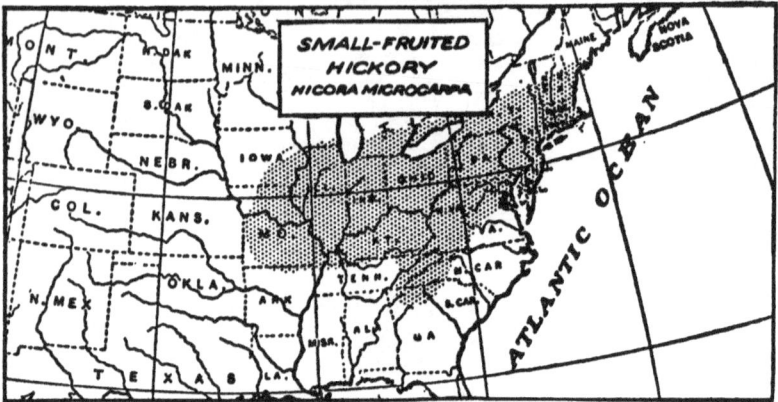

SMALL-FRUITED HICKORY.   (*Hicoria microcarpa*)

A small forest tree up to 90 feet high; considered by some variety of the Pignut; leaves 4 to 7 inches long; it has a small nut *free from angles*; otherwise much like Pignut.

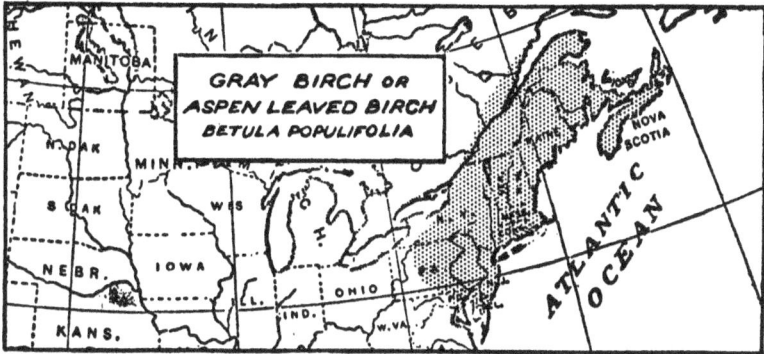

4. BETULACEÆ — BIRCH FAMILY

GRAY BIRCH OR ASPEN-LEAVED BIRCH. (*Betula populifolia*)

A small tree found on dry and poor soil; rarely 50 feet high. Wood soft, close-grained, not strong, splits in drying, useless for weather or ground work. A cubic foot weighs 36 lbs. Leaves 2 to 3 inches long. It has a black triangular scar at each armpit.

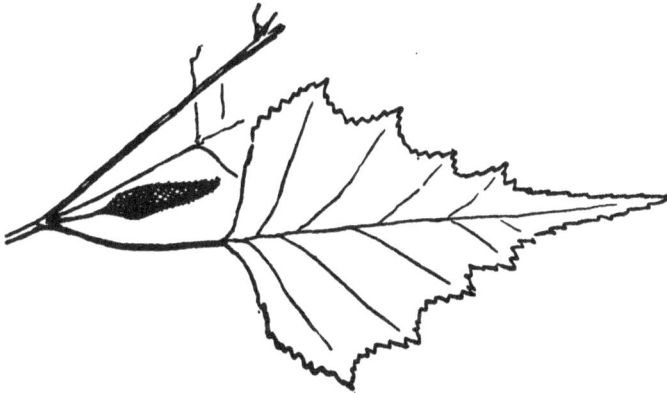

WHITE, CANOE OR PAPER BIRCH. (*Betula papyrifera*)

A tall forest tree up to 80 feet high; the source of bark for canoes, etc. One of the most important trees in the northern forest. Besides canoes, wigwams, vessels and paper from its bark, it furnishes syrup from its sap and the inner bark is used as an emergency food. Every novice rediscovers for himself that the outer bark is highly inflammable as well as waterproof, and ideal for fire-lighting. Though so much like the Gray birch, it is larger, whiter, and without the ugly black scars at each limb. The timber is much the same, but this weighs 37 lbs. Its leaf and catkin distinguish it; the former are 2 to 3 inches long.

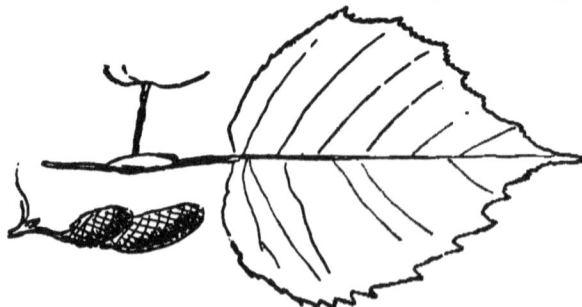

The woodman's fire in *Two Little Savages* was made thus:

> "First a curl of birch bark as dry as it can be,
> Next some sticks of soft wood dead but on the tree;
> Last of all, some pine knots to make the kittle foam,
> An' thar's a fire to make ye think yer sittin' right at home."

This is the noblest of the Birches, the white queen of the woods — the source of food, drink, transport and lodging to those who dwell in the forest; the most bountiful provider of all the trees.

Its sap yields a delicious syrup which has in it a healing balm for the lungs.

Its innermost bark is dried in famine time and powdered to a flour that has some nourishing power.

Its wood furnishes the rims for snowshoes, the frills and fuzzes of its outer bark are the best of fire kindlers, and the timber of the trunk has the rare property of burning whether green or dry.

Its catkins and buds form a favorite food of the partridge which is the choicest of game.

But the outer bark-skin, the famous birch bark, is its finest contribution to man's needs.

The broad sheets of this vegetable rawhide ripped off when the weather is warm and especially when the sap is moving —are tough, light, strong, pliant, absolutely waterproof, almost imperishable in the weather; free from insects, assailable only by fire. It roofs the settler's shack and the forest Indian wigwam, it is the "tin" of the woods and supplies pails, pots, pans, cups, spoons, boxes—under its protecting power the matches are safe and dry, and split very thin, as is easily done, it is the writing paper of the woods, flat, light, smooth, waterproof, tinted and scented; no daughter of the King has ever a more exquisite sheet to sanctify the thoughts committed to its care.

But the crowning glory of the Birch is this — it furnishes the indispensable substance for the bark canoe, whose making is the highest industrial exploit of the Indian life. It would be hard to imagine anything more beautifully made, of and for the life of the Northern woods, buildable, reparable, and usable from the Atlantic to the Pacific, in all the vast region of temperate America — the canoe whose father was the Red mind and whose mother was the birch, is one of the priceless gifts of America to the world. We may use man-made fabrics for the skin, we may substitute unlovely foreign substance for the ribs, or dangerous copper nails for the binding of spruce roots — but the original shape, the lines, the structural ribs, the lipper-turning prow, the roller-riding stern and the forward propulsion of the ever personal paddle, the buoyancy, the wonderful lightness for overland transport, the reparableness by woodland stuffs — these are the things first born of the birch canoe and for these it will be remembered and treasured until man's need of travel on the little waters has reached its final end.

RED BIRCH OR RIVER BIRCH.   (*Betula nigra*)

A tall forest tree of wet banks; up to 90 feet high.   Known by its red-brown scaly bark, of birch-bark style, and its red twigs.   Its wood is light-colored, strong, close-grained, light.   A cubic foot weighs 36 lbs.   Leaves 1½ to 3 inches long.

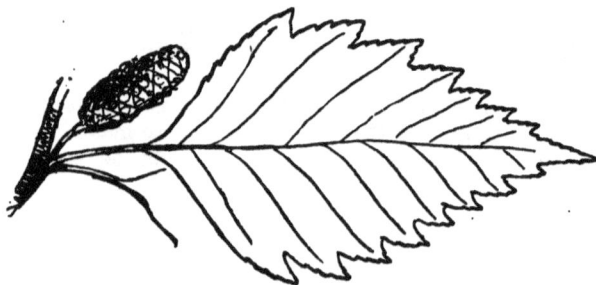

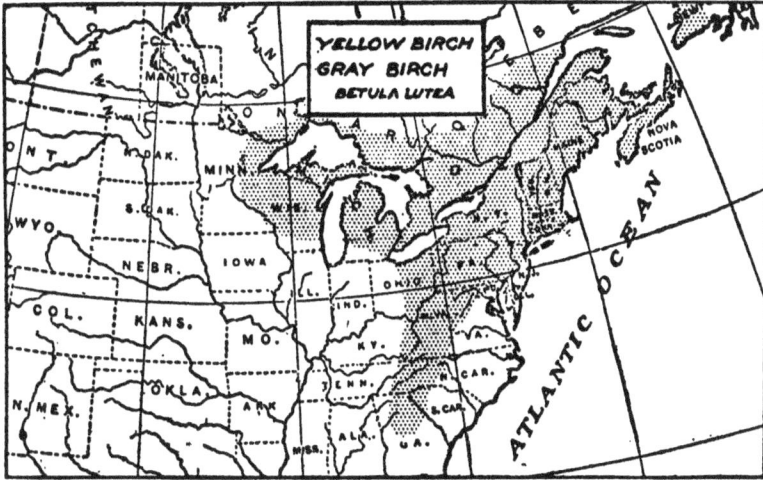

YELLOW BIRCH, GRAY BIRCH.   (*Betula lutea*)

A forest tree, of 30 to 50 feet height.   Bark obviously birch, but shaggy and gray or dull yellow.   Wood as in the others, but reddish. A cubic foot weighs 41 lbs.   Leaves 3 to 4 inches long.

BLACK, CHERRY, SWEET OR MAHOGANY BIRCH
BETULA LENTA

BLACK, CHERRY, SWEET OR MAHOGANY BIRCH. (*Betula lenta*)

The largest of the birches; a great tree, in Northern forests, up to 80 feet high. The bark is little birchy, rather like that of cherry, very dark, and aromatic. Wood dark, hard, clear-grained, very strong; used much for imitating mahogany. A cubic foot weighs 47 lbs. Noted for its sweet, aromatic twigs which made into tea are a fine tonic.

"A decoction of the bark with copperas, is used for coloring woolen a beautiful and permanent drab, bordering on wine color." (*Emerson.*)

Leaves 2½ to 6 inches long. An oil in the bark is very good for sprains and rheumatism.

ALDER OR SMOOTH ALDER, TAG ALDER. (*Alnus serrulata*)

This is the bush so well known in thickets along the Northern streams. It is usually under 20 feet in height, but sometimes reaches 40. Its wood is soft, light brown and useless, a cubic foot weighs 29 lbs. Leaves 3 to 5 inches long. Its inner bark yields a rich orange dye. A tea made of the leaves is a valuable tonic and skin wash for pimples. In wet places or on hillsides.

Besides *serrulata* there are four alders in our limits, the Mountain Alder (*A. alnobetula*) with downy twigs, smooth leaves broad but pointed,

nut with wings; the Speckled Alder (*A. incana*) leaves downy beneath; the European Alder (*A. glutinosa*) with broad, rounded double-toothed leaves; (this often becomes a tall tree) and the Seaside Alder (*A. maritima*) known by its long narrow leaves.

IRONWOOD, HARD-HACK, LEVERWOOD, BEETLE-WOOD OR HOP HORN-
BEAM. (*Ostrya Virginiana*)

A small tree; 20 to 30, rarely 50 feet high; named for its hardness and its hop-like fruit. Bark, furrowed. Wood, tough close-grained, unsplittable. One of the strongest, heaviest and hardest of timbers. A cubic foot weighs over 51 lbs. That is, it comes near to Shagbark Hickory in weight and perhaps goes beyond it in strength and hardness. Leaves 3 to 5 inches long. Fruit 1½ to 2½ inches long.

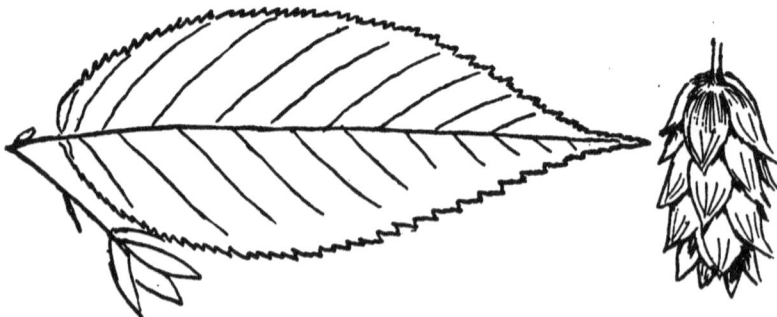

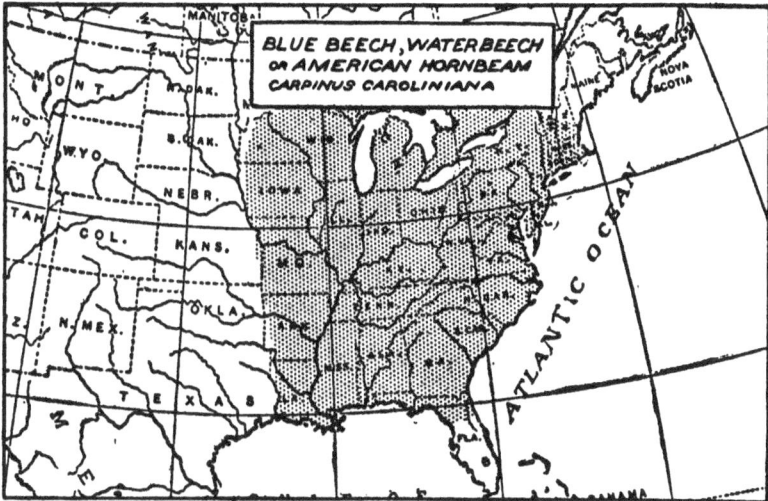

BLUE BEECH, WATER BEECH OR AMERICAN HORNBEAM. (*Carpinus caroliniana*)

A small tree, 10 to 25 feet, rarely 40 feet high; bark, smooth. Wood hard close-grained, very strong; much like Ironwood, but lighter. A cubic foot weighs 45 lbs. Leaves 3 to 4 inches long.

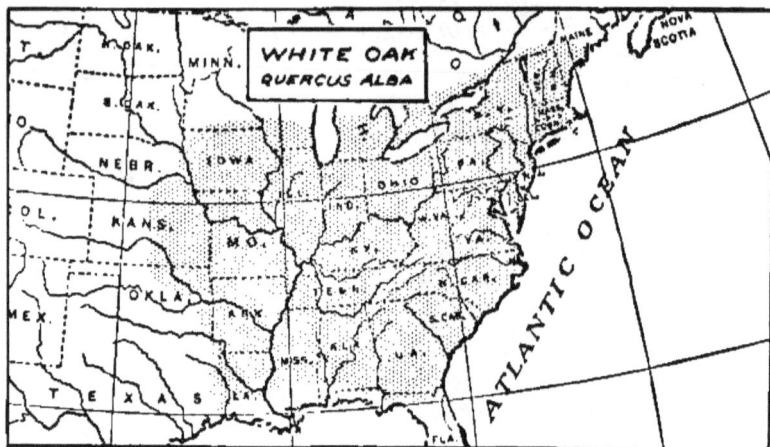

WHITE OAK
QUERCUS ALBA

5. FAGACEÆ — BEECH FAMILY

WHITE OAK. (*Quercus alba*)

A grand forest tree; over 100 feet up to 150 feet high. The finest and most valuable of our oaks. The one perfect timber for shipbuilders, farmers and house furnishers. Its wood is pale, strong, tough, fine-grained, durable and heavy. A cubic foot weighs 46 lbs. I found that when green it weighed 68 lbs. to the cubic foot and of course sank in water like a stone. Called white from pale color of bark and wood. Leaves 5 to 9 inches long. Many of them hang all winter though dead so the White Oak contributes a little to the golden glow of the snowy woods, though not to the extent of the Black Oak. Its acorns ripen in one season. They are sweet and nutritious and eagerly sought after by every creature in the woods from bluejays, wild ducks, mice and deer to squirrels and schoolboys.

There can be little doubt that at least three out of five nut trees were planted by squirrels, chiefly the gray squirrel. All through autumn before snow falls the industrial Bannertail Gray works to bury for future use the choicest nuts he finds on the ground; ignoring the coarse and bitter, he makes sure of the sweet and delicate. Those that are not so disposed of, are usually eaten by deer, bears and other wild things. The various oaks have long competed for the squirrels' attention to their product. The Bur Oak acorn attracted by its size, Chestnut Oak by its split-ability and the White Oak by the sweetness. For a time the White

Oak fared well, for it furnished indeed the most delectable of our nuts, but now it is in an evil case. Largely through the growing scarceness of the gray squirrel the White Oak, the most valuable of its group, is no longer planted throughout its range. Its edibility is now a menace to its life, for it lies exposed and all things eagerly devour it while the other acorns lie untouched and we are now threatened with the extermination of this our noblest oak, the one that chiefly gave value to our hardwood forests, partly at least I believe through the near-extinction of the gray squirrel, its unwitting protector. The connection between these two creatures is so intimate that their ranges coincide exactly throughout the length and breadth of the land.

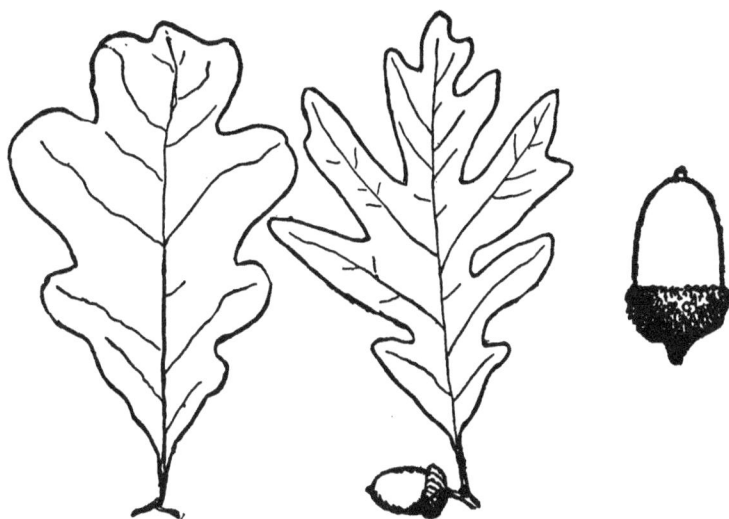

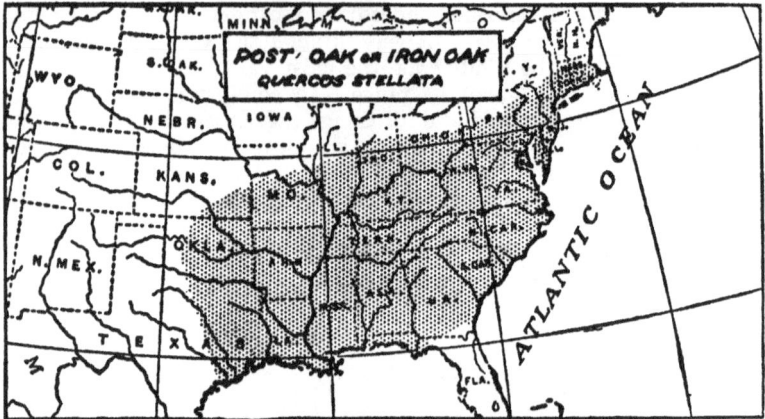

POST OAK, OR IRON OAK. (*Quercus stellata*)

A smaller tree, rarely 100 feet high; of very hard wood, durable; used for posts, etc. A cubic foot weighs 52 lbs.; that is, the same as Shagbark Hickory. Leaves 5 to 8 inches long. Acorns ripen in one season.

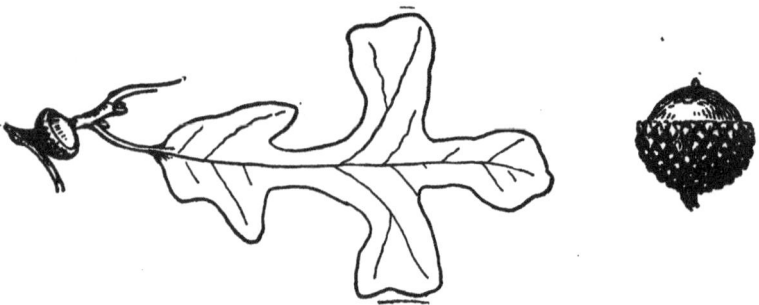

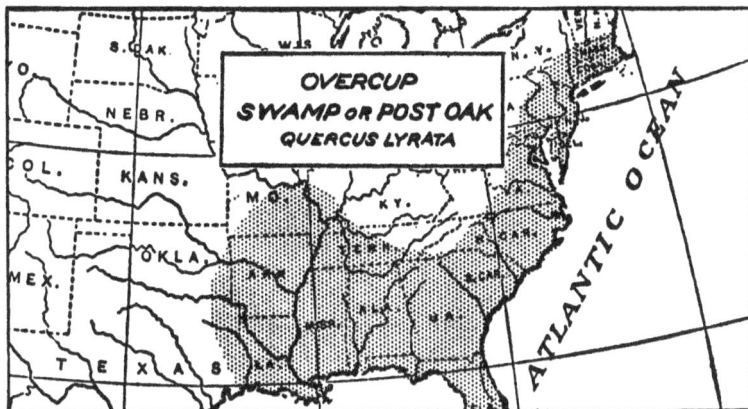

OVERCUP, SWAMP OR POST OAK. (*Quercus lyrata*)

A large tree up to 100 feet high. Wood very strong and durable; a cubic foot weighs 52 lbs. Noted for the cup covering the acorn. Leaves 6 to 8 inches long.

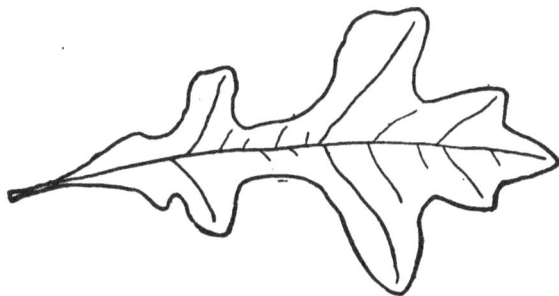

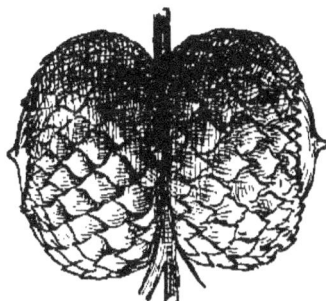

Bur Oak, Cork Bark or Mossy Cup. (*Quercus macrocarpa*)

A large forest tree, up to 160 feet high; known by its enormous acorns and the *corky ridges* on the twigs.   The cork of commerce is the bark of an oak found in Spain and it's not surprising to find a cork bark in our own land.   The leaves though greatly varied are alike in having two deep bays one on each side near the middle dividing the leaf nearly to the midrib so that the type is as given below; they are 4 to 8 inches long. The acorns ripen in one season.   The wood is like that of most oaks, and lasts well next the ground.   A cubic foot weighs 46 lbs.

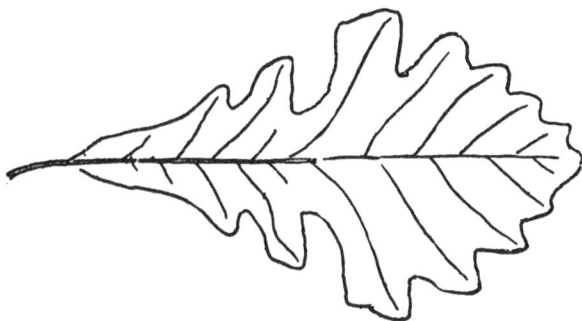

Leaf and acorn of Bur Oak
(acorn life size)

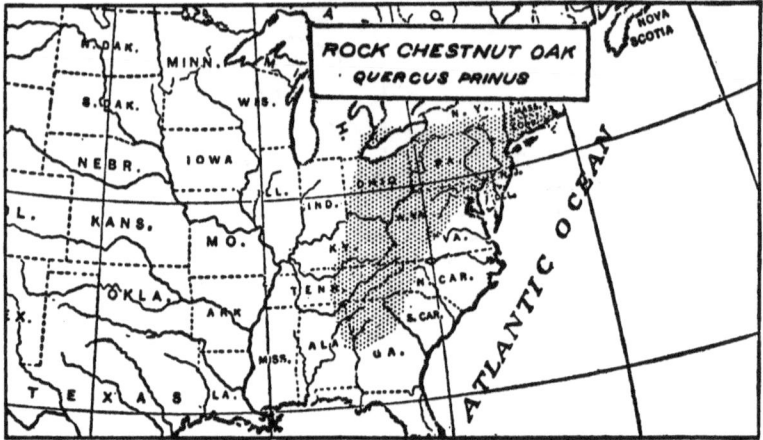

ROCK CHESTNUT OAK. (*Quercus prinus*)

A good sized tree; up to 100 feet high. Wood as usual. A cubic foot weighs 47 lbs. Its acorns are immense, 1¼ to 1½ inches long, and ripen in one season. Leaves 5 to 10 inches long.

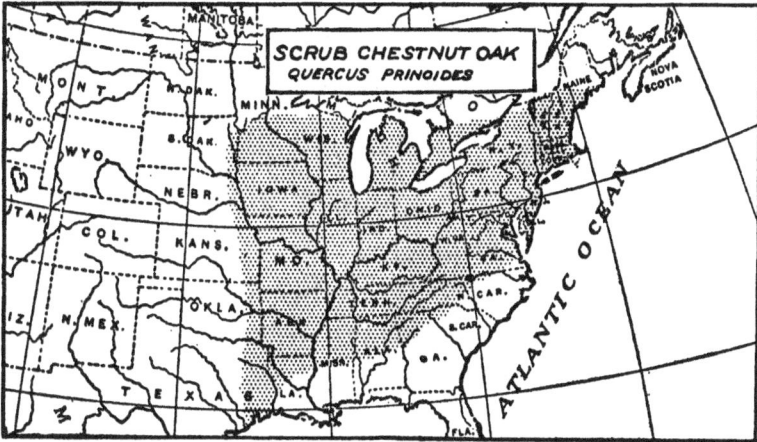

SCRUB CHESTNUT OAK. (*Quercus prinoides*)

A mere shrub, 2 to 15 feet high. Close akin to the preceding. Leaves 2½ to 5 inches long. Found in dry sandy and poor soil.

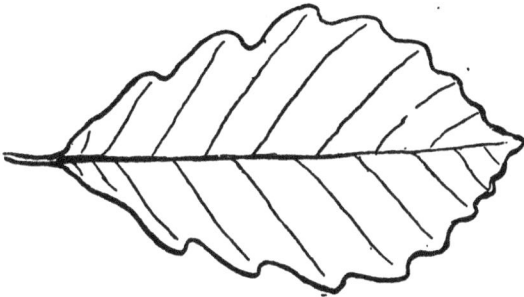

YELLOW OAK, CHESTNUT OAK, SCRUB OAK or CHINQUAPIN
QUERCUS MUHLENBERGII

YELLOW OAK, CHESTNUT OAK OR CHINQUAPIN SCRUB OAK.
(*Quercus Muhlenbergii*)

A great forest tree; up to 160 feet high; wood as usual, but the heaviest of all, when dry; a cubic foot weighs 54 lbs; when green, it is heavier than water, and sinks at once. It is much like the Chestnut Oak but its leaves are narrower, more sharply saw-edged and its acorns much smaller, about half the size. Its acorns ripen in one season. Leaves 4 to 6 inches long.

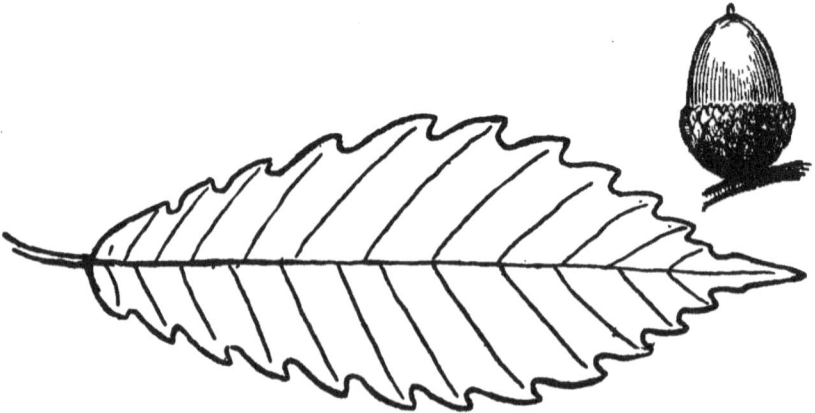

SWAMP WHITE OAK. (*Quercus bicolor*)

A fine forest tree in swampy land; up to 110 feet high. Wood as in preceding species, but a cubic foot weighs only 48 lbs. It has the leaf of a White Oak, the bark of a Black. Its smaller branches have the bark rough and loose giving a shaggy appearance to the tree. Its acorns ripen in one season and as in all the annual fruiting oaks its wood is durable next the ground.

RED OAK. (*Quercus rubra*)

RED OAK. (*Quercus rubra*)

A fine forest tree, 70 to 80, or even 140, feet high. Wood reddish-brown. Sapwood darker. Hard, strong, coarse-grained, heavy. A cubic foot weighs 41 lbs. It checks warps and does not stand for weather or ground work. The acorn takes two seasons to ripen. Apparently all those oaks whose nuts take *two* seasons to ripen have wood that soon rots. The low flat shape of the cup is distinctive; in fact it has no cup, it has a saucer; leaves 4 to 8 inches long.

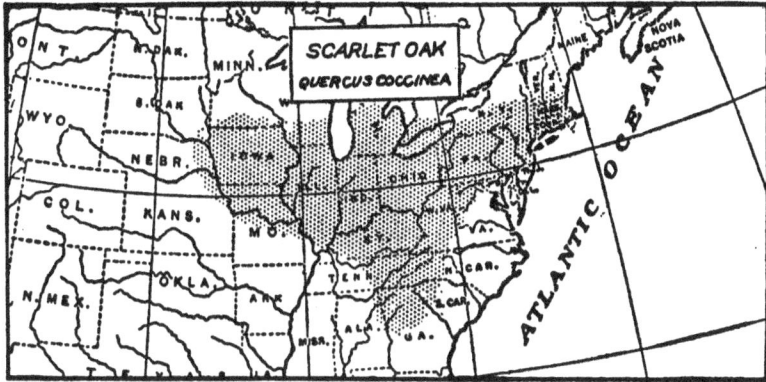

SCARLET OAK. (*Quercus coccinea*)

Seventy to 80 or even 160 feet high. Scarlet from its spring and autumn foliage color. The leaves are a little like those of the Black Oak, but are frond-like with three or four deep, nearly even, cuts on each side: The acorns of this can be easily matched among those of the Black Oak, but the kernel of the Scarlet is white, that of the Black is yellow; they take two seasons to ripen. Wood much as in Red Oak but weighs 46 lbs. per cubic foot. Leaves 4 to 8 inches long.

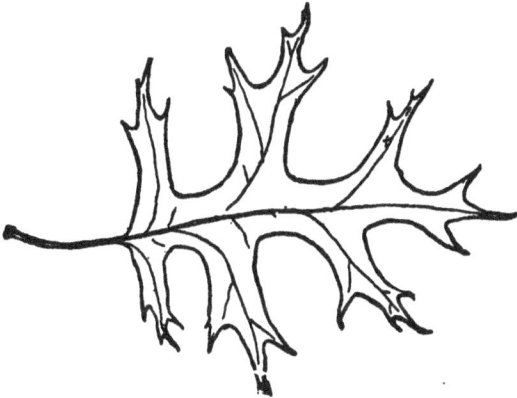

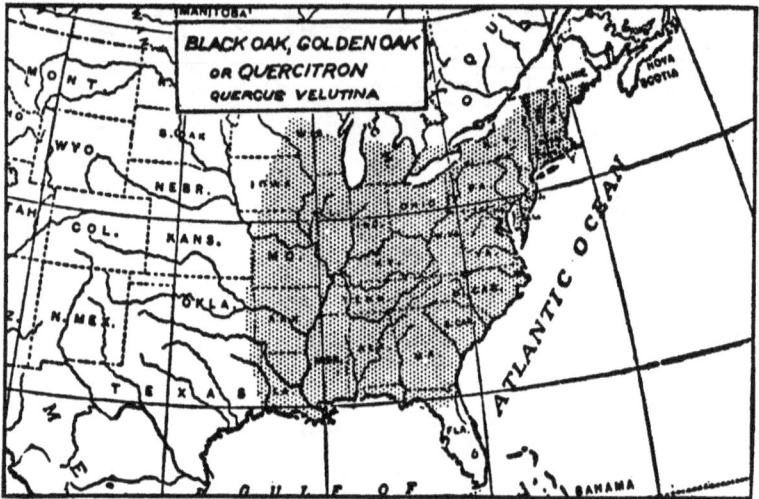

BLACK OAK, GOLDEN OAK OR QUERCITRON. (*Quercus velutina*)

Seventy to 80 or even 150 feet high. The outer bark is very rough, bumpy and blackish; inner bark yellow. This yields a yellow dye called *quercitron*. The leaf is of the Scarlet Oak style, but has uneven cuts and usually a large solid area in the outer half. The wood is hard, coarse-grained, checks, and does not stand for weather or ground work. A cubic foot weighs 44 lbs. The acorns take two seasons to ripen. Taking the White Oak acorn as a standard of white, that is a yellowish-white, the acorn of the present when cut open is a distinct golden yellow. As in all oaks the leaves vary greatly, look for the

type not the exact portrait among the illustrations; they are 4 to 6 inches long.

One of the wonderful things about this oak is the persistence of its leaves. Though dead and faded they cling in numbers to the tree all winter; their exquisite old gold is one of the artist's joys and the glory of the winter landscape. This with its bright yellow inner bark, its bright yellow nut and its yellow brown winter foliage amply entitle it to be called "golden oak."

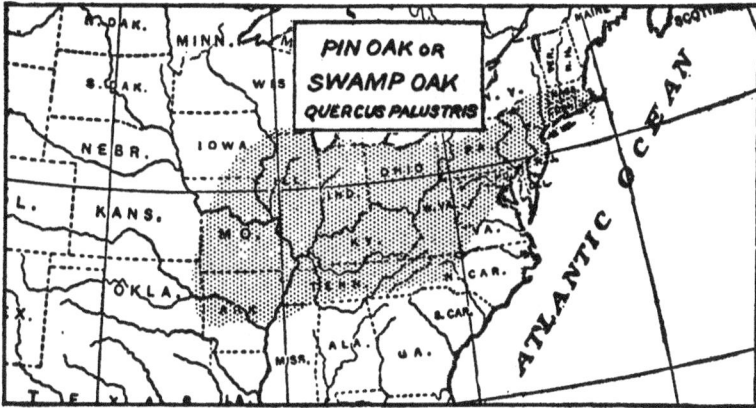

PIN OAK OR SWAMP OAK. (*Quercus palustris*)

Fifty to 70 or even 120 feet high, in swampy land. Wood hard, coarse-grained, very strong and tough; the Pin Oak is more happily named than most of its kin, first the numerous short straight branches in the lower trunk, make it seem stuck full of large pins, next, each point of its leaves has a pin on it, in each armpit of the midrib below is a tiny velvet pin cushion and finally and chiefly this exceptionally tough wood

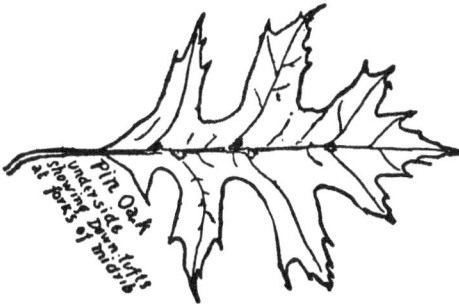

was the best available for making the pins in frame barns. In Wyndygoul Park I cut a Pin Oak that was 110 feet high and 32 inches across
the stump and yet had but 76 rings of annual growth. Will not stand
exposure next to ground. A cubic foot weighs 34 lbs. Its acorns take
two seasons to ripen. Leaves 4 to 6 inches long. In moist woods and
along swamp edges.

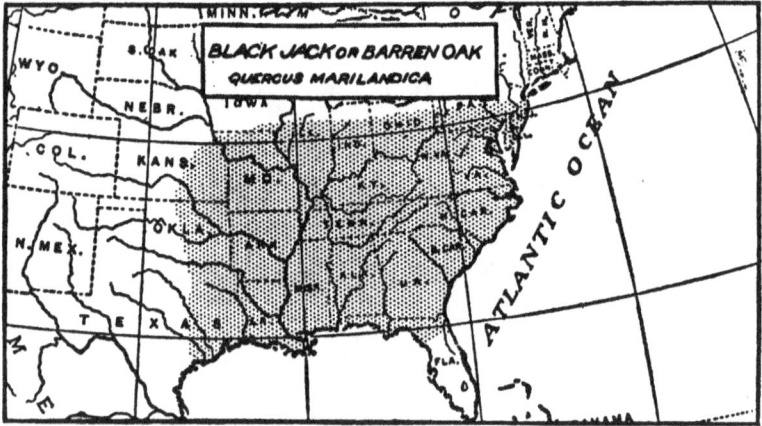

BLACK JACK OR BARREN OAK. (*Quercus marilandica*)

A small tree seldom up to 60 feet high. An unimportant tree of barren
wastes. Leaves 3 to 5 lobed downy below, bristle-tipped and 3 to 7
inches long; acorns take two seasons to ripen. Wood hard and dark,
not durable. A cubic foot weighs 46 lbs.

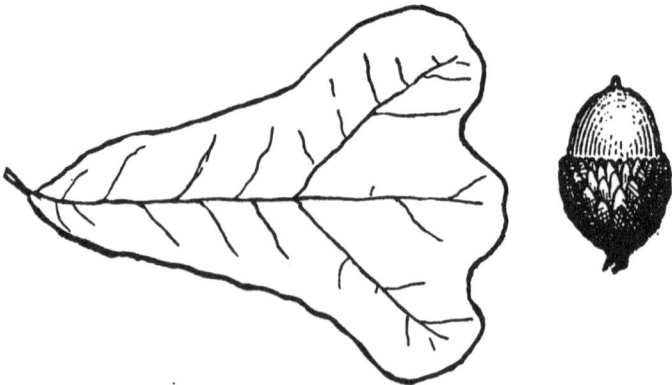

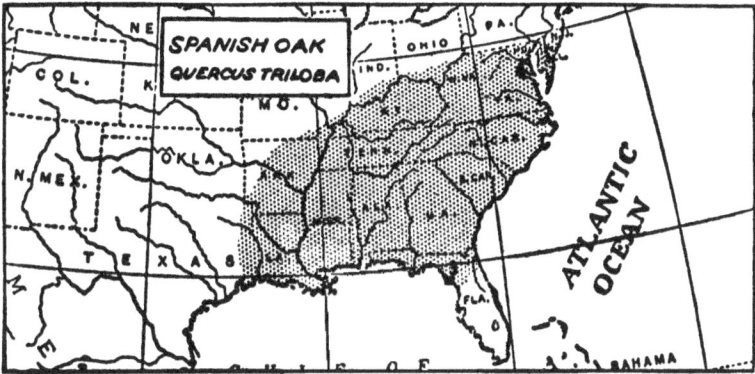

SPANISH OAK. (*Quercus triloba*)

A large tree up to 100 feet occasionally. Found on dry soil. Leaves bristle-tipped, 5 to 7 inches long, with 3 to 7 lobes. The acorns do not ripen till the second year so we may expect the wood to be undurable. A cubic foot of it weighs 43 lbs.

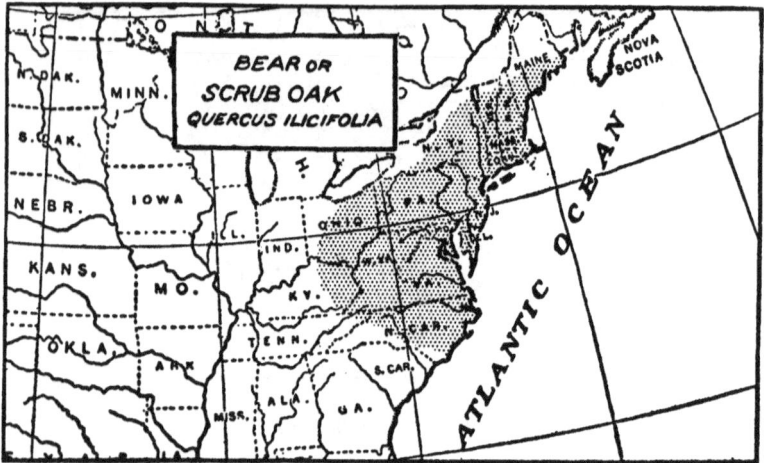

BEAR OR SCRUB OAK. (*Quercus ilicifolia*)

An insignificant tree rarely 25 feet high.   Often forming dense thickets, on poor sandy or rocky soil.   The leaves are bristle-tipped, 2 to 5 inches long.   The acorns ripen in the second season and are so bitter that nobody cares who gets them.   The bears were least squeamish so were welcome to the crop hence one of the names.

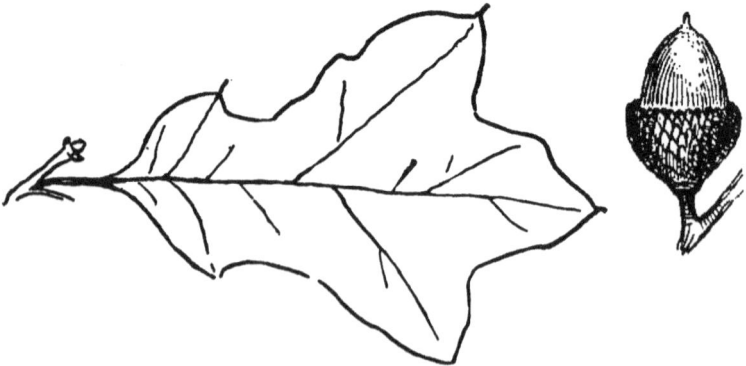

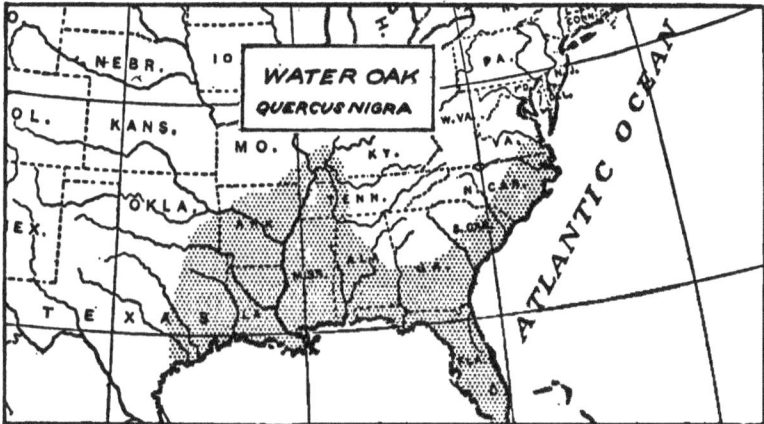

WATER OAK. (*Quercus nigra*)

A middle-sized tree, rarely 80 feet high, found chiefly along streams and swamps. Leaves 1½ to 3 inches long; 1 to 3 lobed at the end. Wood hard and strong, a cubic foot weighs 45 lbs. The acorns ripen in the second season so look out for its timber. This leaf has tufts of hair in the armpits of the veins beneath.

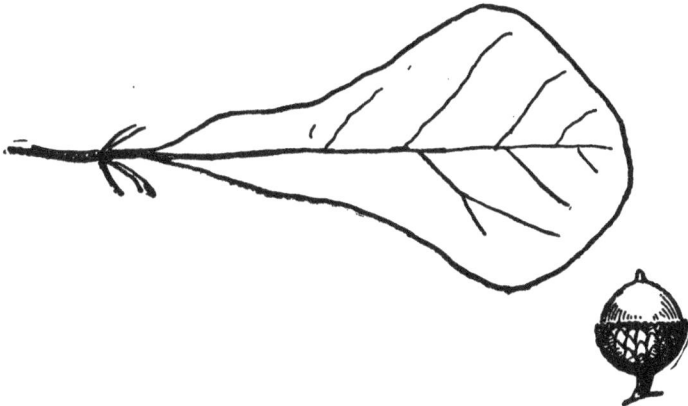

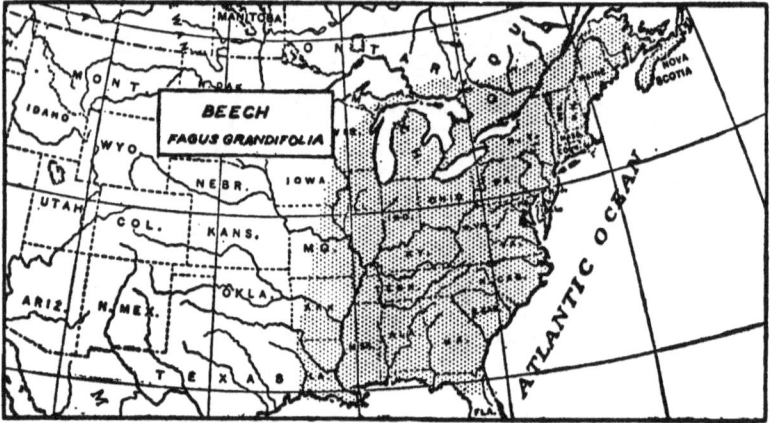

BEECH.   (*Fagus grandifolia*)

In all North America there is but one species of beech.   It is a noble forest tree, 70 to 80, and occasionally 120 feet high; readily distinguished by its unfurrowed ashy gray bark.   Wood hard, strong, tough, close-grained, pale, heavy.   Leaves 3 to 4 inches long.   A cubic foot weighs 43 lbs.   It shares with Hickory and Sugar Maple the honor of being a perfect firewood.

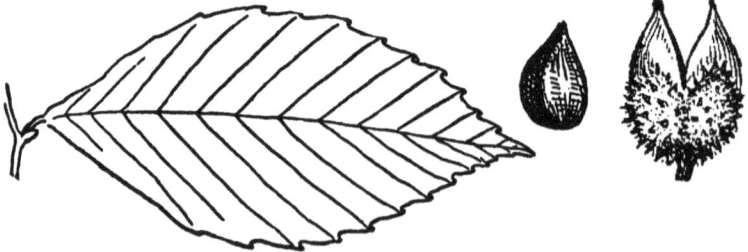

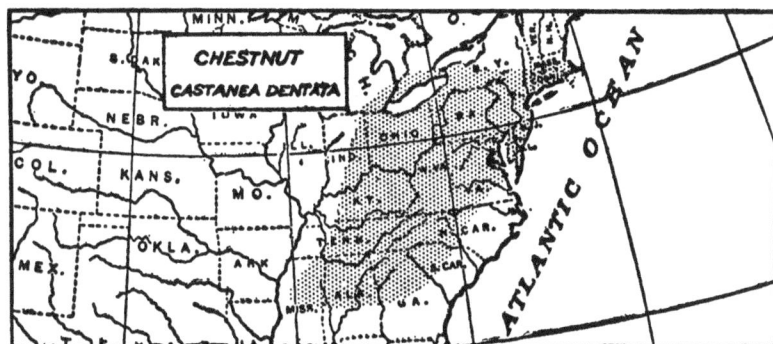

CHESTNUT. (*Castanea dentata*)

A noble tree, 60 to 80 or even 100 feet high. Whenever you see something kept under lock and key, bars and bolts, guarded and double guarded, you may be sure it is very precious, greatly coveted — the nut of this tree is hung high aloft, wrapped in a silk wrapper which is enclosed in a case of sole leather, which again is packed in a mass of shock-absorbing vermin-proof pulp, sealed up in a waterproof iron-wood safe and finally cased in a vegetable porcupine of spines, almost impregnable.

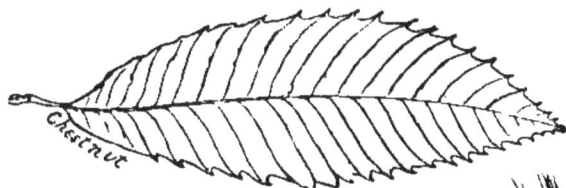

There is no other nut so protected; there is no nut in our woods to compare with it as food. Wood, brown, soft, easily worked, coarse, too easily split, very durable as posts or other exposed work, altogether a most valuable timber, the present plague that threatens to wipe it out is a fungus probably from abroad. There is no known remedy. A cubic foot of the wood weighs 28 lbs. Leaves 6 to 8 inches long.

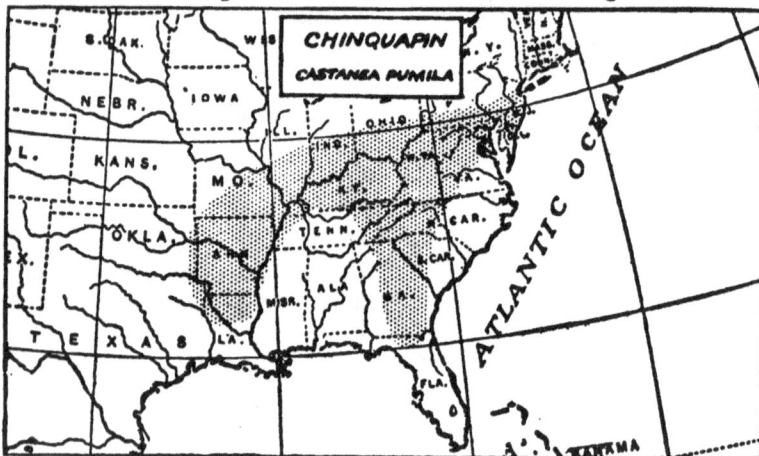

CHINQUAPIN. (*Castanea pumila*)

A small tree, rarely 45 feet high, with the general character of the common Chestnut. It is much smaller in all ways. Its leaves are 3

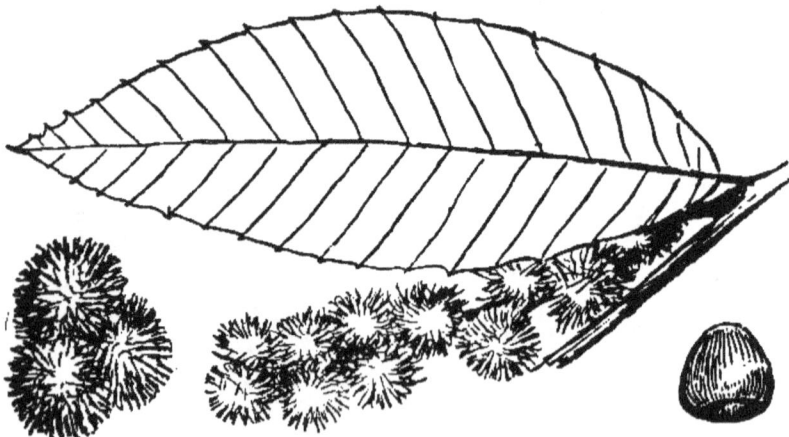

to 6 inches long; its burs less than half the size of *dentata*. Its wood is similar but darker and heavier, a cubic foot weighing 37 lbs. These two complete the list of chestnuts native to the Northeastern States.

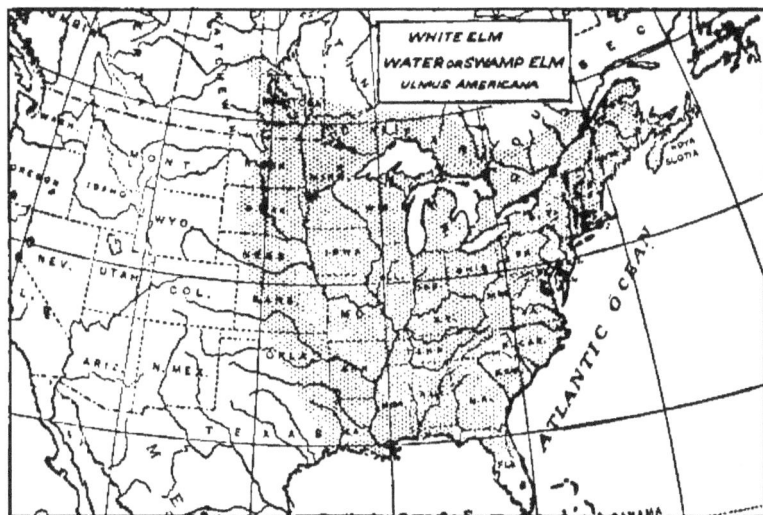

6. ULMACEÆ — ELM FAMILY

WHITE ELM, WATER OR SWAMP ELM. (*Ulmus Americana*)

A tall splendid forest tree; commonly 100, occasionally 120 feet high. Wood reddish-brown; hard, strong, tough, very hard to split. This furnished the material of the hubs in O. W. Holmes's "One Hoss Shay."

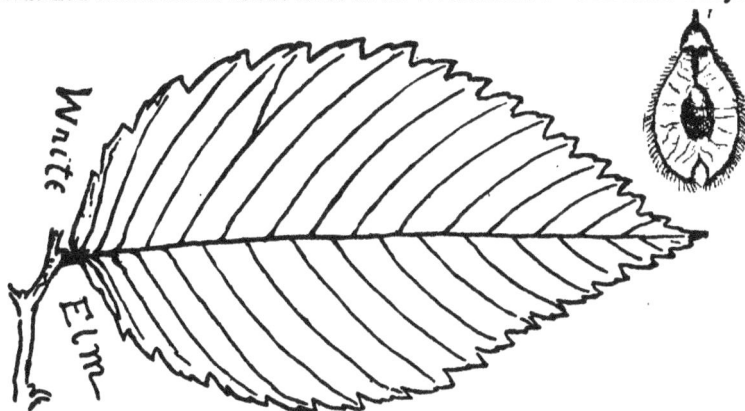

It is coarse, heavy; fairly good firewood, but sparks badly. A cubic foot weighs 41 lbs. Soon rots near the ground. Leaves 2 to 5 inches long. Flowers in early spring before leafing. Seeds ripe in May. Common in most parks.

SLIPPERY ELM, MOOSE OR RED ELM. (*Ulmus fulva*)

Smaller than White Elm, maximum height about 70 feet. Wood dark, reddish; hard, close, tough, strong; durable next the ground; heavy; a cubic foot weighs 43 lbs. Its leaves are *larger and rougher* than those of the former. Four to 8 inches long, and its buds are hairy, not smooth. The seeds ripen in early spring when the leaves are half grown; they were a favorite spring food of the Passenger Pigeon. Chiefly noted for its mucilaginous buds, inner bark and seeds, which are eaten or in decoction used as a cough-remedy. This is a valuable specific in all sorts of membranous irritation: for the hard cough or bowel trouble, drink it; for sores apply it in poultice form. It can never do harm and always does some good.

The inner bark of this Elm contains a great quantity of mucilage, and is a favorite popular prescription, in many parts of the country, for dysentery and affections of the chest.

"It is much to be regretted that the Slippery Elm has become so rare. The inner bark is one of the best applications known for affections of the throat and lungs. Flour prepared from the bark by drying perfectly and grinding, and mixed with milk, like arrow-root, is a wholesome and nutritious food for infants and invalids." (*Emerson.*)

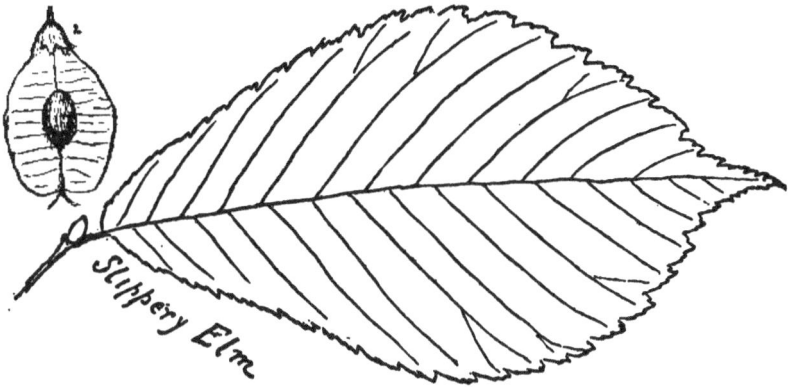

1. American Elm   3. Cork Elm
2. Slippery Elm   4. Wahoo

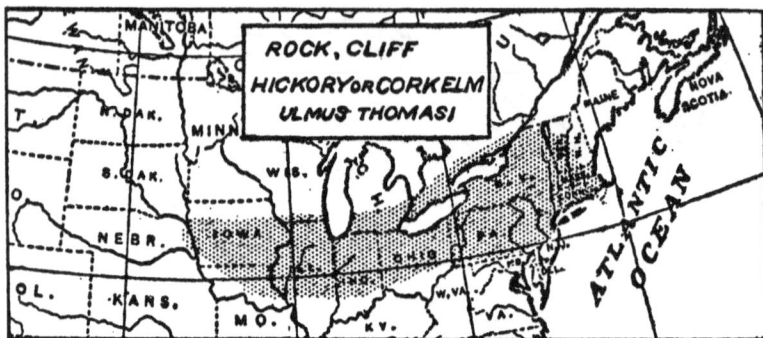

ROCK, CLIFF, HICKORY OR CORK ELM. (*Ulmus Thomasi*)

A tall forest tree on dry or rocky uplands; occasionally 100 feet high. Wood pale, reddish-brown; hard, close, strong, tough and heavy. A cubic foot weighs 45 lbs. It lasts a long time next the ground. It is regularly marked with corky ridges on the two-year-old branches, which give it a shaggy appearance. Its leaves are 2 to 5 inches long. "It possesses all the good qualities of the family, and none of the bad ones." (*Keeler.*)

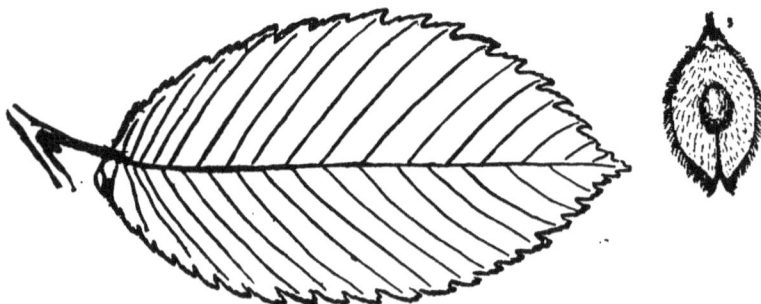

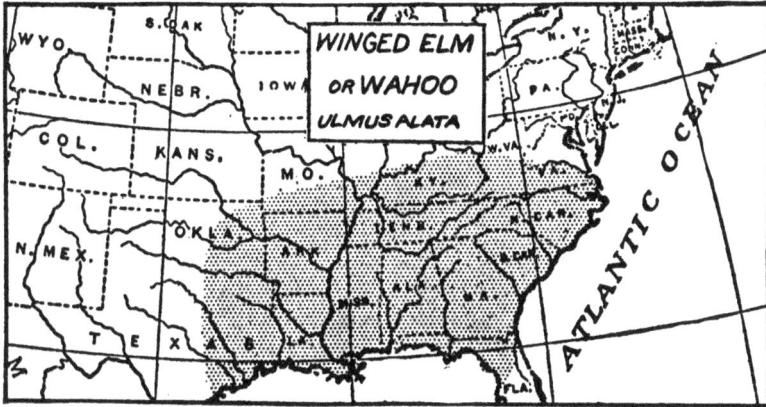

WINGED ELM OR WAHOO. (*Ulmus alata*)

A small tree, up to 50 feet high. Remarkable for the flat corky wings on most of the branches. The wood is hard, weak and brown. A cubic foot weighs 47 lbs. Its leaves are 1 to 3 inches long.

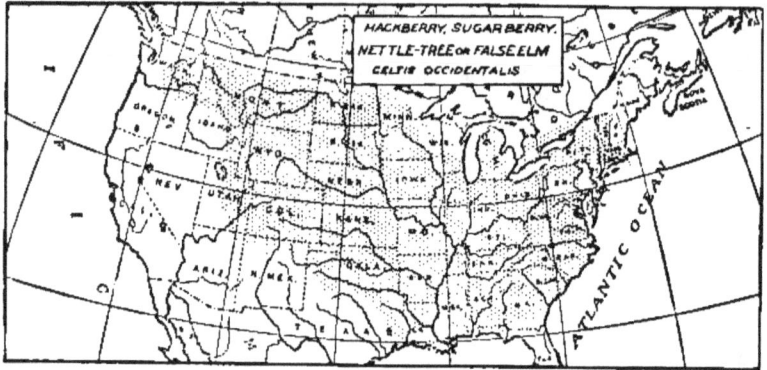

HACKBERRY, SUGARBERRY, NETTLE TREE OR FALSE ELM. (*Celtis occidentalis*)

A tall slender tree, 50 feet, rarely 100 feet high. Wood soft, pale, coarse, a cubic foot weighs 45 lbs. Leaves 2 to 6 inches long. Its style is somewhat elm-like, but it has small dark purple berries, each with a large stone like a cherry pit. The wood is "used for the shafts and axletrees of carriages, the naves of wheels, and for musical instruments. The root is used for dyeing yellow; the bark for tanning; and an oil is expressed from the stones of the fruit." (*Emerson.*) In dry soil.

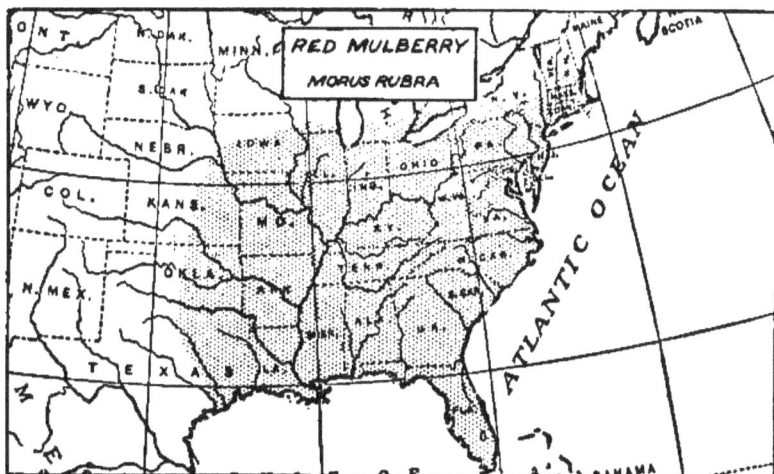

7. MORACEÆ — MULBERRY FAMILY

RED MULBERRY. (*Morus rubra*)

A fine forest tree up to 65 feet high; wood, pale yellow, soft, weak but durable; a cubic foot weighs 37 lbs.; berries 1½ inches long, dark purple red, delicious. Leaves 3 to 5 inches long. In rich soil.

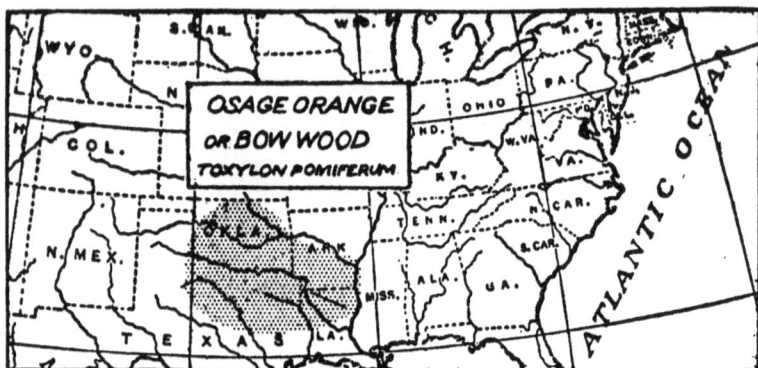

OSAGE ORANGE, BOIS D'ARC, BODARC OR BOW-WOOD.   (*Toxylon pomiferum*)

A small tree, rarely 60 feet high. Originally from the middle Mississippi Valley, now widely introduced as a hedge tree. Famous for supplying the best bows in America east of the Rockies. Wood is bright orange; very hard, elastic, enduring and heavy. Leaves 3 to 6 inches long. A cubic foot weighs 48 lbs.

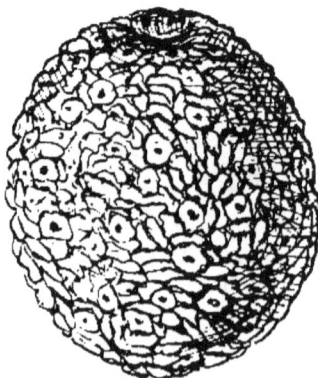

Orange, ¼ of life size

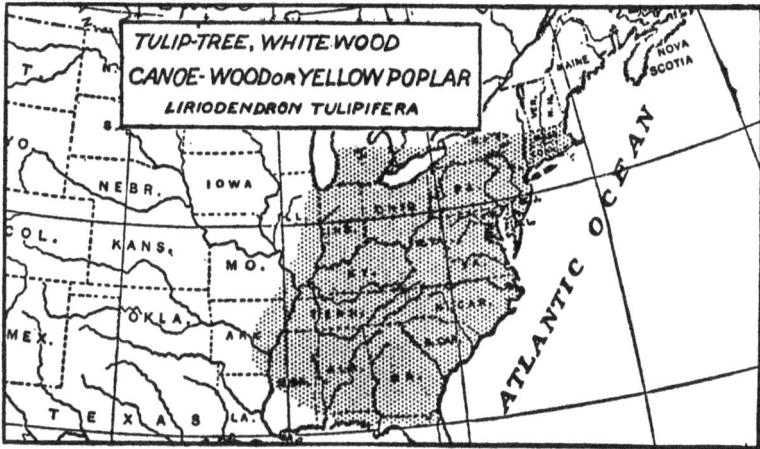

TULIP-TREE, WHITE-WOOD
CANOE-WOODoRYELLOW POPLAR
LIRIODENDRON TULIPIFERA

8. MAGNOLIACEÆ — MAGNOLIA FAMILY

TULIP TREE, WHITE-WOOD, CANOE WOOD OR YELLOW POPLAR. (*Liriodendron Tulipifera*)

One of the noblest forest trees, ordinarily 100 feet, and sometimes 150 feet high. Noted for its splendid clean straight column; readily known by leaf, 3 to 6 inches long, and its tulip-like flower. Wood soft, straight-grained, brittle, yellow, and very light; much used where a broad sheet easily worked is needed but will not stand exposure to the weather; is poor fuel; a dry cubic foot weighs 26 lbs.

Makes a good dugout canoe, hence Indian name, "canoe wood" (*Keeler*). The inner bark and root bark either as dry powder or as "tea" are powerful tonics and especially good for worms.

Every tree like every man must decide for itself — will it live in the alluring forest and struggle to the top where alone is sunlight or give up the fight and content itself with the shade — or leave this delectable land of loam and water and be satisfied with the waste and barren plains that are not desirable.

The Tulip is one of those that believe there is plenty of room at the top and its towering trunk is one of the noblest in the woods that shed their leaves. The Laurel and Swamp Magnolia are among the shadow dwellers; and the Scrub Oaks and the Red Sumacs are among those that have lost in the big fight and are content with that which others do not covet.

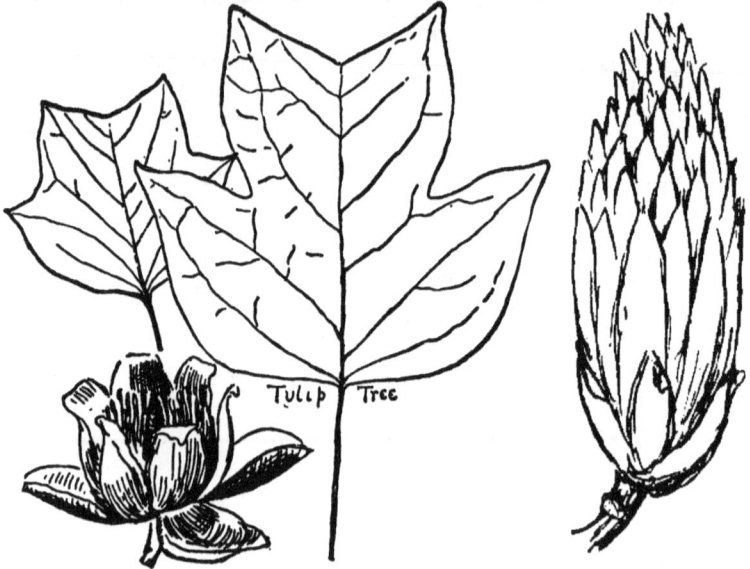

Tulip Tree

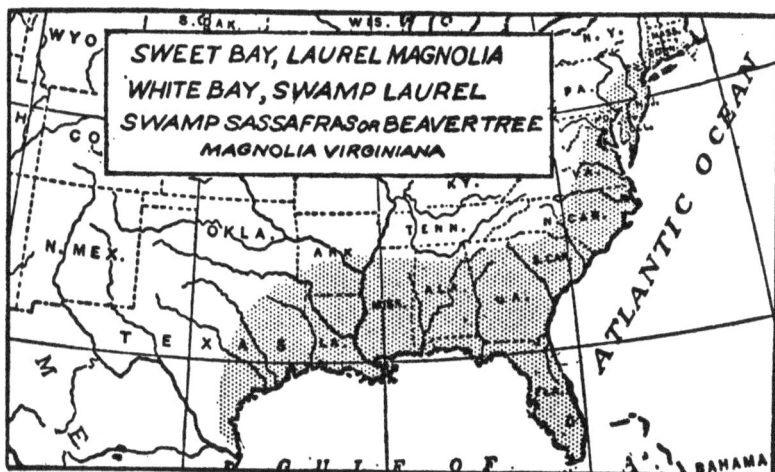

SWEET BAY, LAUREL MAGNOLIA, WHITE BAY, SWAMP LAUREL, SWAMP
SASSAFRAS OR BEAVER TREE. (*Magnolia virginiana*)

A small tree 15 to 70 feet high, nearly evergreen, noted for being a
favorite with the Beaver. "Its fleshy roots were eagerly eaten by the
Beavers, who considered them such a dainty that they could be caught
in traps baited with them. Michaux recites that the wood was used by
the beavers in constructing their dams and houses in preference to any
other." (*Keeler.*)

The wood weighs 31 lbs. to the cubic foot. The heart wood is reddish-
brown, the sap wood nearly white. The leaves are 3 to 6 inches long,
dark shiny green above, faintly downy below. Fruit cone 1½ to 2
inches high.

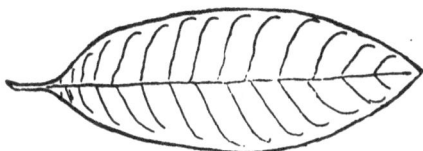

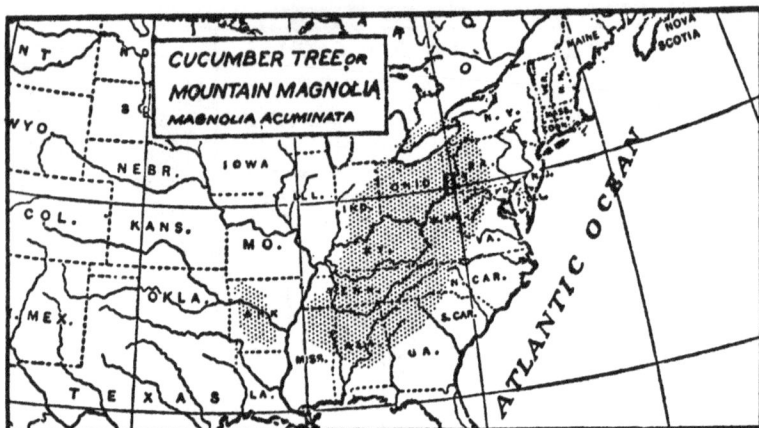

CUCUMBER TREE OR MOUNTAIN MAGNOLIA. (*Magnolia acuminata*)

A fair-sized forest tree 60 to 90 feet high. The wood weighs 29 lbs. to the cubic foot. The leaves are light green, faintly downy below, 2 to 12 inches long. Fruit cone 3 to 4 inches high.

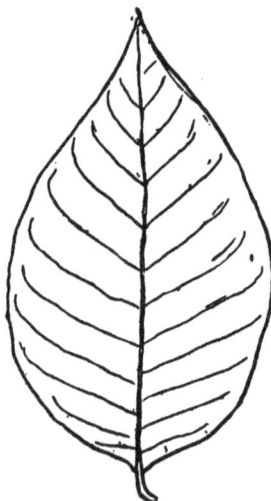

¼ life size

SPICE BUSH, FEVER BUSH
WILD ALL SPICE, BENJAMIN BUSH
BENZOIN ODORIFERUM

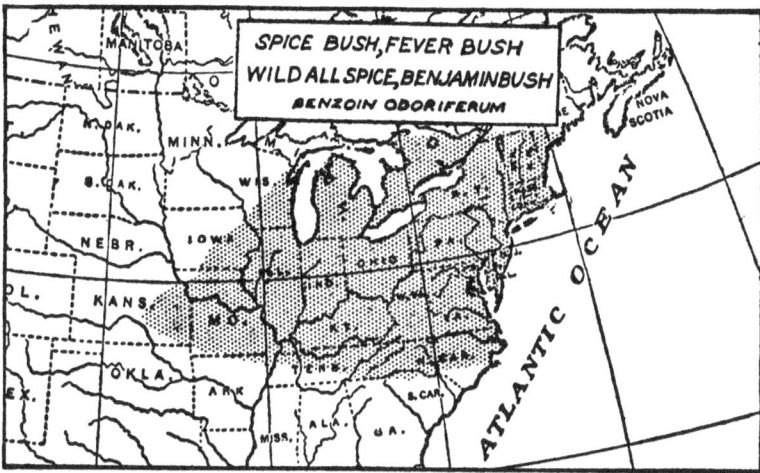

9. LAURACEÆ — LAUREL FAMILY

SPICE BUSH, FEVER BUSH, WILD ALLSPICE, BANJAMIN BUSH.   (*Benzoin odoriferum*)

A small bush rarely 20 feet high.   In moist woods; berries red; leaves 2 to 5 inches long.   A tea made of its twigs was a good old remedy for chills and fever.

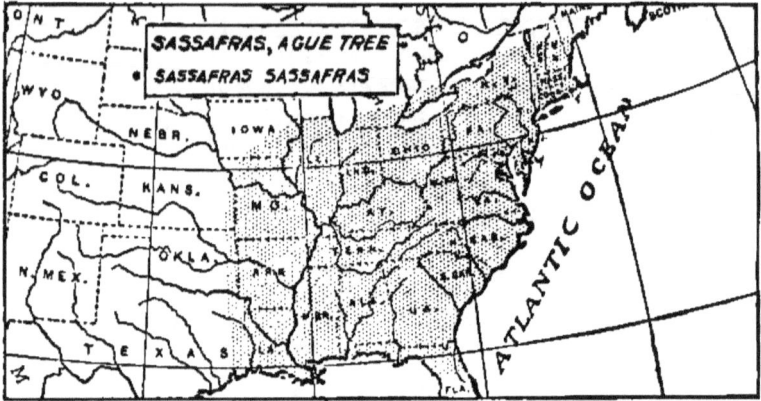

SASSAFRAS, AGUE TREE.   (*Sassafras Sassafras*)

Usually a small tree of dry sandy soil, but reaching 125 feet high in favorable regions.   Its wood is dull orange, soft, weak, coarse, brittle, and light.   A cubic foot weighs 31 lbs.   Very durable next the ground. Leaves 4 to 7 inches long.   Noted for its aromatic odor.

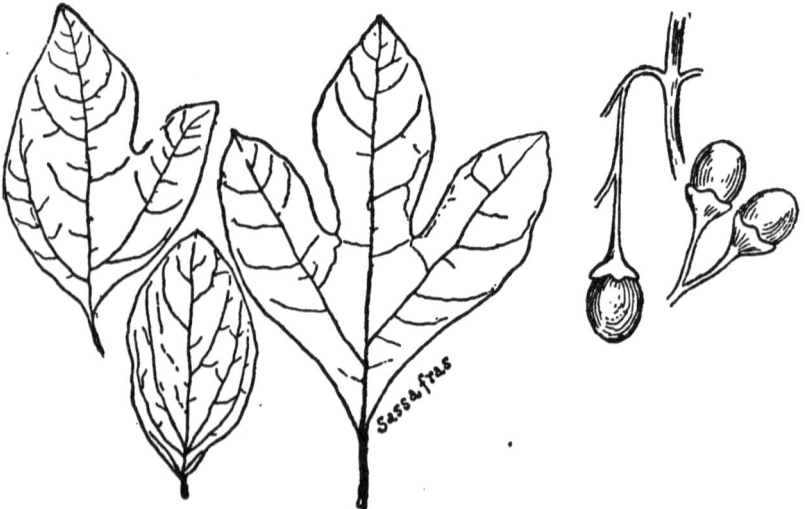

"In the Southwestern States the dried leaves are much used as an ingredient in soups, for which they are well adapted by the abundance

of mucilage they contain. For this purpose the mature green leaves are dried and powdered, the stringy portions being separated, and are sifted and preserved for use. This preparation mixed with soups, give them a ropy consistence, and a peculiar flavor, much relished by those accustomed to it. To such soups are given the names *gombo file* and *gombo zab.* (P. 321.)

"A decoction of the bark is said to communicate to wool a durable orange color." (P. 322) (*Emerson*).

Tea made of the bark is also a fine warming stimulant and sweater. Its roots are used in the manufacture of root-beer.

WITCH HAZEL, WINTER BLOOM OR SNAPPING HAZEL NUT
HAMAMELIS VIRGINIANA

10. HAMAMELIDACEÆ — WITCH-HAZEL FAMILY

WITCH-HAZEL WINTER BLOOM OR SNAPPING HAZEL NUT. (*Hamamelis virginiana*)

A small tree 10 to 15 feet high, usually with many leaning stems from one root. Noted for its blooming in the fall, flowers of golden threads, the nuts explode when ripe throwing the seeds a dozen feet. A snuff made of the dry leaves stops nosebleed at once, or indeed any bleeding when

locally applied.   A decoction or tea of the bark gives relief to inflamma-
tion of the eye or skin.

> Witch hazel blossoms in the fall
> To cure the chills and fever all.
> *(Two Little Savages.)*

A forked twig of this furnished the favorite divining rod whence the
name.   Leaves 4 to 6 inches long.

11. ALTINGIACEÆ — SWEET GUM FAMILY

SWEET GUM, STAR-LEAVED OR RED GUM, BILSTED, ALLIGATOR TREE OR LIQUIDAMBAR. (*Liquidambar Styraciflua*)

A tall tree up to 150 feet high of low, moist woods, remarkable for the corky ridges on its bark, and the unsplittable nature of its weak, warping, perishable timber. Heart-wood reddish-brown, sap white; heavy, weighing 37 lbs. to cubic foot. Leaves 3 to 5 inches long.

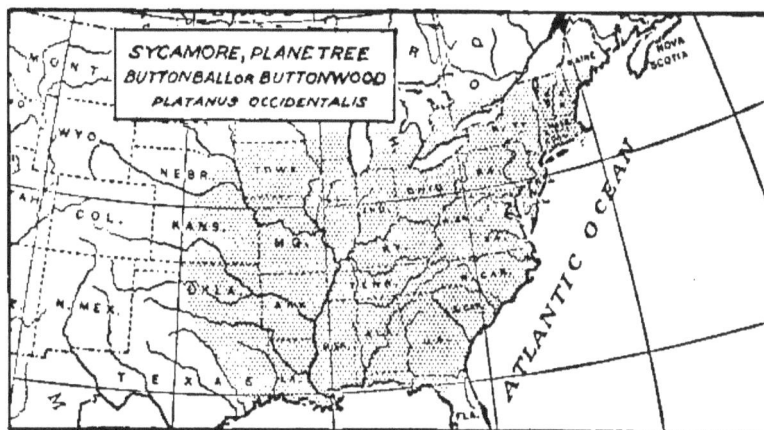

12. PLATANACEÆ — PLANE TREE FAMILY

SYCAMORE, PLANE TREE, BUTTONBALL OR BUTTONWOOD. (*Platanus occidentalis*)

One of the largest of our trees; up to 140 feet high; commonly hollow. Wood, light brownish, weak; hard to split; heavy for its strength. A cubic foot weighs 35 lbs. Little use for weather work. Famous for

shedding its bark as well as its leaves.   Leaves 4 to 9 inches long. Canada to the Gulf.

When a tree is a mere sapling, the bark is thin and soft; it stretches each year with the annual growth of the trunk.   But it becomes thicker and harder with age and then it cracks with the expansion of the trunk. This process continues each year till the segments of the first coat are widely separated by gaping fissures.   This is well seen in the Elm, and each of the bark ridges shows the annual layers, from the widely separated outer one to the united inmost one.

But some trees, notably the Sycamore, burst their bark, yet do not retain the fragments.   These are dropped each year, hence the smooth green surface of the trunk, hence also its success as a tree of grimy cities, for it has an annual cleaning of the skin and thus throws off mischievous accumulations that would kill a tree that retained its bark indefinitely.

The Shagbark Hickory will be remembered as a halfway shedder.

CHOKE CHERRY
PADUS VIRGINIANA

13. AMYGDALACEÆ — PLUM FAMILY

CHOKE CHERRY. (*Padus virginiana*)

A bush 2 to 19 feet high in the North. A tall tree in the Mississippi Valley. Wood, pale, hard, close-grained, and heavy. A cubic foot weighs 43 lbs. Leaves 2 to 4 inches long, the marginal teeth divaricate or outcurved. Noted for its astringent fruit. Leaf broader, fruit smaller than in Black Cherry.

Teeth enlarged

BLACK CHERRY, CABINET OF RUM CHERRY.   (*Padus serotina*)

A fine tree, even in Canada; 60 to 70 or even 90 feet high.   The source of many excellent remedies, chiefly pectoral.   Tea of the bark (roots preferred) is a powerful tonic for lungs and bowels; also good as

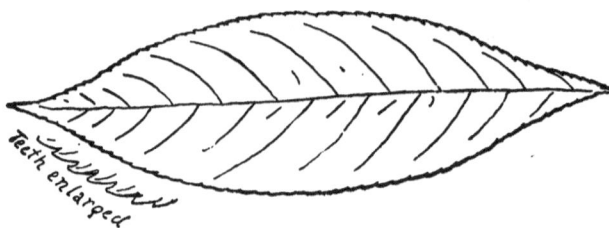

a skin wash for sores. The leaves when half wilted are poisonous to cattle. The wood is light-brown or red, strong, close-grained; much in demand for cabinet work; light. A cubic foot weighs 36 lbs. Leaves 5 inches long, the marginal teeth incurved.

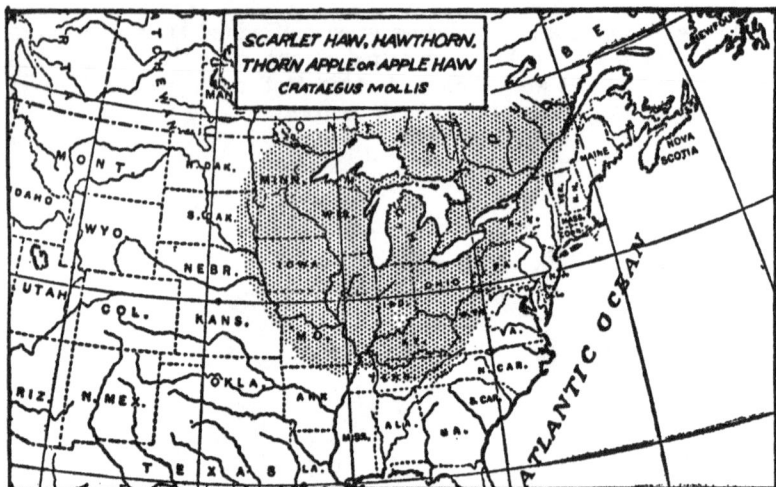

SCARLET HAW, HAWTHORN,
THORN APPLE or APPLE HAW
CRATAEGUS MOLLIS

14. MALACEÆ — APPLE FAMILY

SCARLET HAW, HAWTHORN, THORN APPLE OR APPLE HAW.   (*Cratægus mollis*)

A small tree, 10 to 20, rarely 30 feet high. Wood hard and heavy. A cubic foot weighs 50 lbs. Leaves 2 to 4 inches long. Noted for its beautiful deep red fruit, ¾ to 1¼ inches long, round, with pink-yellow flesh, 5 or 6 stones, quite eatable.

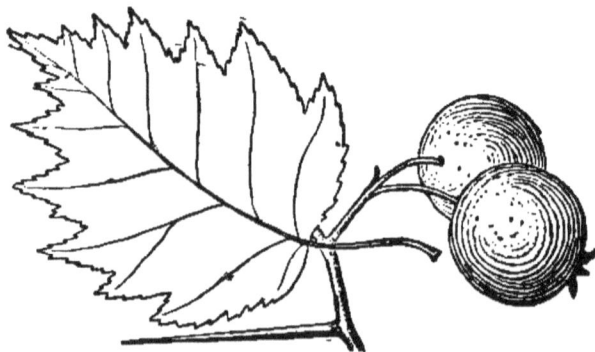

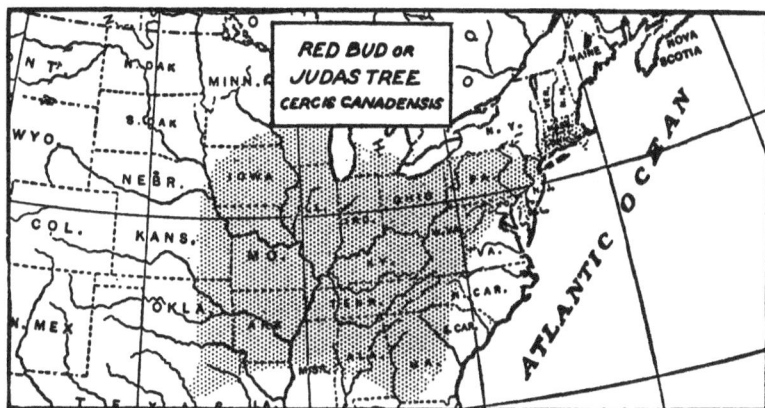

15. CÆSALPINACEÆ — SENNA FAMILY

RED-BUD OR JUDAS TREE. (*Cercis canadensis*)

Small tree of bottom lands, rarely 50 feet high; so called from its abundant spring crop of tiny rosy blossoms, coming before the leaves, the latter 2 to 6 inches broad. "Judas tree" because it blushed when Judas hanged himself on it. (*Keeler.*) Its wood is dark, coarse and heavy.

A cubic foot weighs 40 lbs.

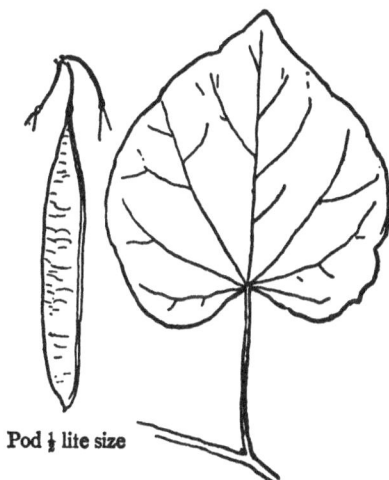

Pod ½ lite size

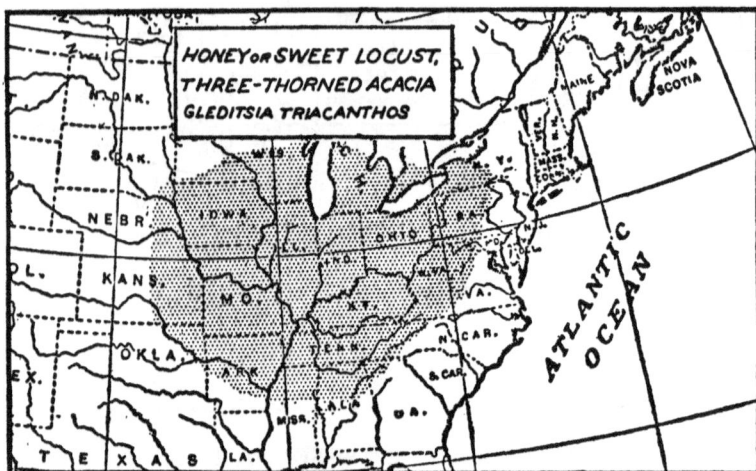

HONEY or SWEET LOCUST,
THREE-THORNED ACACIA
GLEDITSIA TRIACANTHOS

HONEY OR SWEET LOCUST, THREE-THORNED ACACIA.
(*Gleditsia triacanthos*)

A tall tree up to 140 feet high; very thorny. Wood dark, hard, strong, coarse, heavy. A cubic foot weighs 42 lbs. Leaves single or double pinnate; leaflets ¾ to 1¼ inches long. It is very durable as posts, etc. Pods 6 to 12 inches long. So called because of the sweet stuff in which its seeds are packed. Chiefly Mississippi Valley, but common in the East along roadsides.

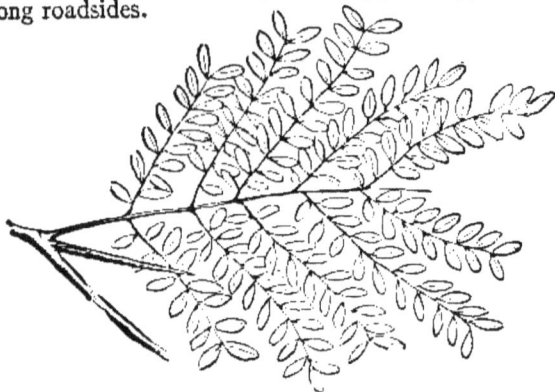

Pod is ½ life size

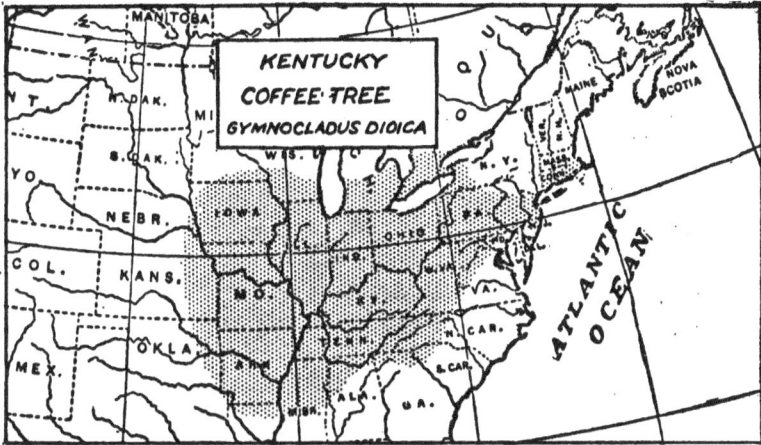

KENTUCKY COFFEE TREE. (*Gymnocladus dioica*)

A tall tree (up to 100 feet), so called because its beans were once used as coffee. Wood is light-colored, coarse-grained strong, and heavy. A cubic foot weighs 43 lbs. Leaves large and bipinnate; leaflets, 7 to 15, and 1 to 3 inches long. It is remarkably durable next the ground, as posts, etc.

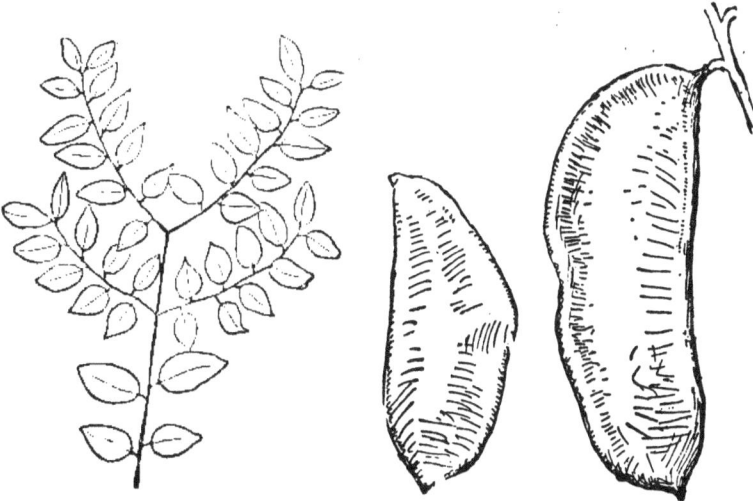

Pods ½ life size

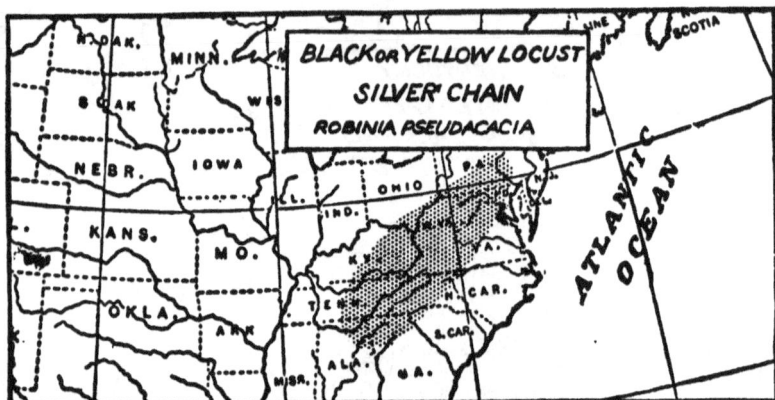

BLACK or YELLOW LOCUST
SILVER CHAIN
ROBINIA PSEUDACACIA

16. FABACEÆ — PEA FAMILY

BLACK OR YELLOW LOCUST, SILVER CHAIN. (*Robinia Pseudacacia*)

A tall forest tree, up to 80 feet high: leaves 8 to 14 inches long; leaflets 9 to 19, 1 to 2 inches long; pods 2 to 4 inches long, 4 to 7 seeded. Wood greenish-brown, very strong and durable; much used for posts; weight 46 lbs. per cubic foot.

"The leaves are used in some parts of Europe, either fresh or cured, as nourishment for horses; the seeds are found very nutritious to fowls.

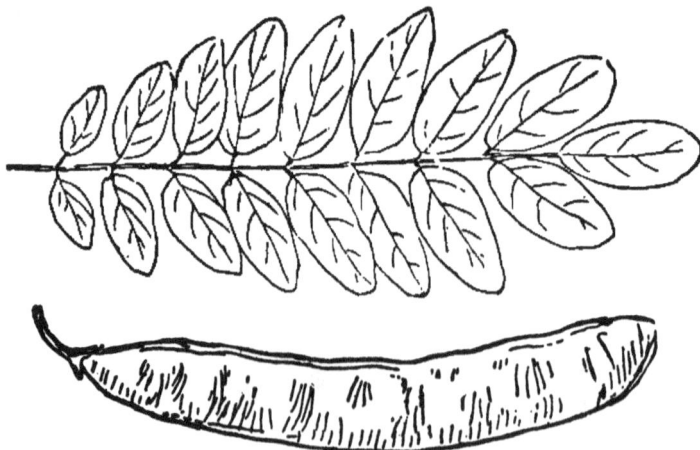

The leaves may be made a substitute for indigo in dyeing blue, and the flowers are used by the Chinese for dyeing yellow." (*Emerson.*)

Pennsylvania to Iowa and South to Georgia and common in the east along roadsides.

STAGHORN or VELVET SUMAC

VINEGAR TREE

RHUS HIRTA

17. ANACARDIACEÆ — SUMAC FAMILY

STAGHORN OR VELVET SUMAC, VINEGAR TREE. (*Rhus hirta*)

A small tree 10 to 40 feet high. Noted for its red velvety berries in solid bunches and its *velvet clad stem* whence its name. Leaflets 11 to 31 and 2 to 5 inches long; the whole leaf 16 to 24 inches long.

"The berries are also used in dyeing their own color. Kalm says, that the branches boiled with the berries, afford a black, ink-like tincture." (*Emerson.*)

Nova Scotia to British Columbia, south to Florida and west to Arizona.

Somewhat like it but *quite smooth* is the Smooth or Scarlet Sumac. (*R. glabra.*)

Its berries make a safe and pleasant drink for children and tea of almost any part of the tree is a powerful tonic.

Leaves and fruit of
Scarlet Sumac

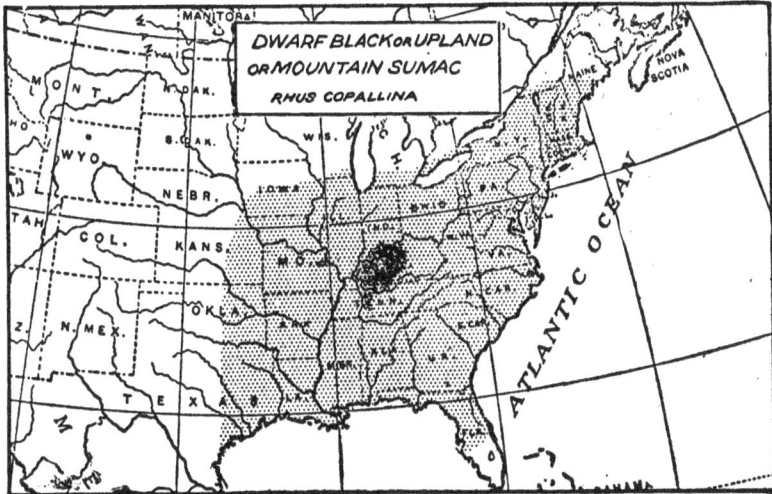

DWARF, BLACK, UPLAND OR MOUNTAIN SUMAC. (*Rhus copallina*)

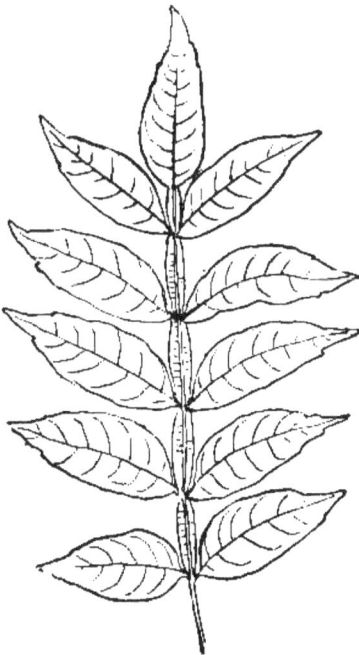

A small tree like the Staghorn; of similar range. Known by the peculiar winged stems of the leaves. Leaves 6 to 12 inches long and leaflets 2 to 4 inches long; number 9 to 21. Dry soil. Maine to Minnesota and south to Florida and Texas.

POISON SUMAC, POISON ELDER.   (*Rhus Vernix*)

A small tree, 15 to 20 up to 25 feet high.   Noted for being the most poisonous tree in the country.   Its active principle is a fixed oil.   This may be removed by washing with an alcoholic solution of sugar of lead; it is a sure cure.   When this remedy is not at hand, wash the parts with water as hot as one can stand, this is also a reliable remedy.   The same remarks apply to Poison Ivy and Poison Oak.   Leaves 6 to 15 inches long; leaflets 7 to 13 in numbers and 2 to 4 inches long.   Timber is light and worthless.   A cubic foot weighs 27 lbs.   Damp woods.

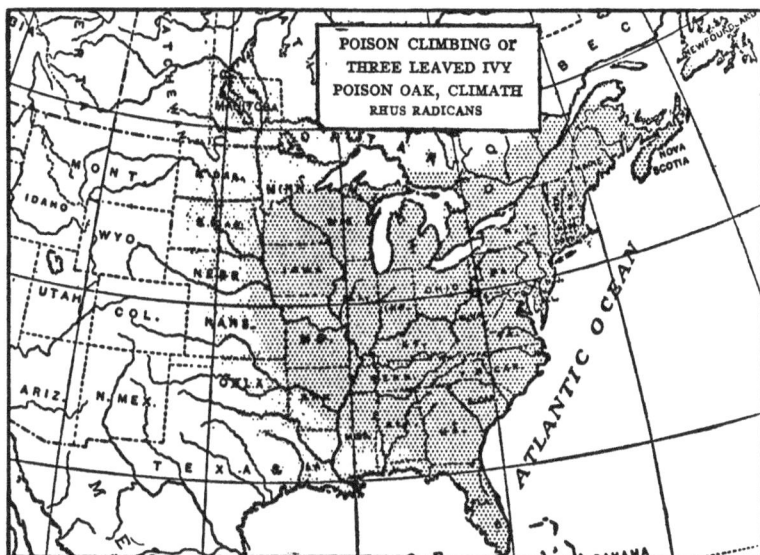

POISON CLIMBING or
THREE LEAVED IVY
POISON OAK, CLIMATH
RHUS RADICANS

Poison Climbing or Three-leaved Ivy.   Poison Oak, Climath.
(*Rhus radicans*)

Though a trailing vine on the ground, on fences or on trees and never itself a tree, the Poison or Three-fingered Ivy should appear here that all may know it. Its poisonous powers are much exaggerated, about three persons out of four are immune and the poison is easily cured as

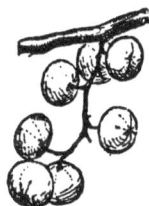

described under Poison Sumac. Its leaflets always three, are 1 to 4 inches long. Its berries are eagerly eaten by birds.

"The juice of this plant is yellowish and milky, becoming black after a short exposure to the air. It has been used as marking ink and on linen is indelible." (*Emerson.*) It grows everywhere in the open being found from Manitoba eastward and Texas northward.

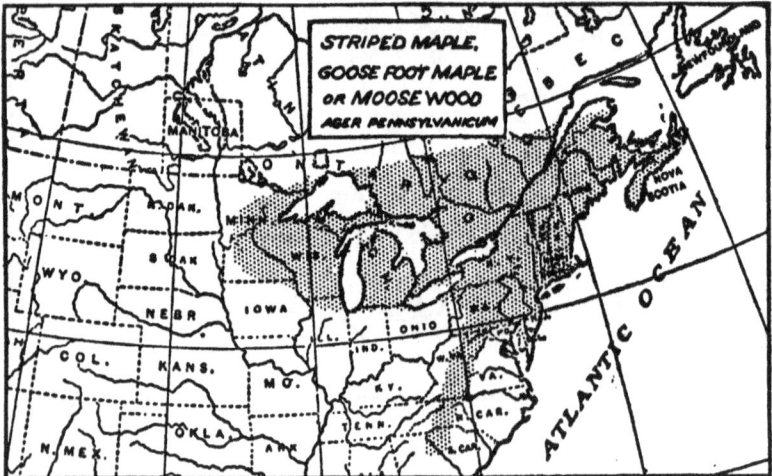

18. ACERACEÆ — MAPLE FAMILY

STRIPED MAPLE, GOOSEFOOT MAPLE OR MOOSEWOOD. (*Acer pennsylvanicum*)

A small tree up to 35 feet high, in tall woods, called "striped" because its small branches have white lines. It is much eaten by the moose. Wood, brown, soft, close-grained, light. Leaves, 5 to 6 inches long. A cubic foot weighs 33 lbs.

MOUNTAIN MAPLE.   (*Acer spicatum*)

A shrub or small tree, rarely 30 feet high.   Wood soft, pale and light, a cubic foot weighs 33 lbs.   Leaves 4 to 5 inches along.

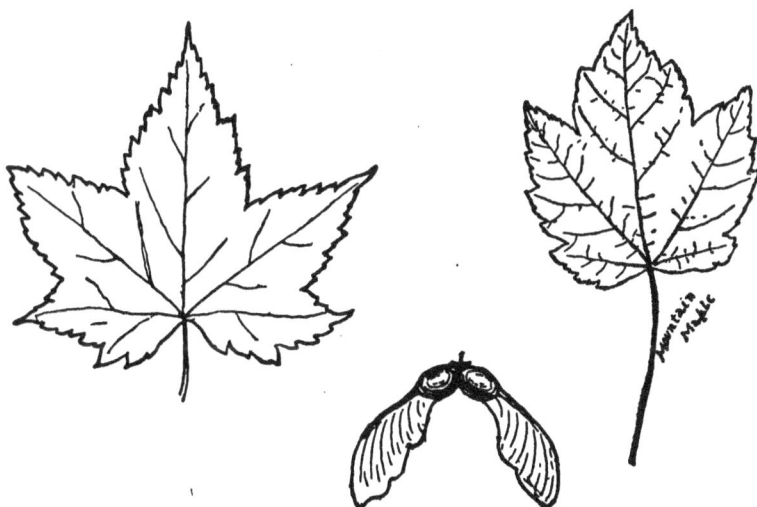

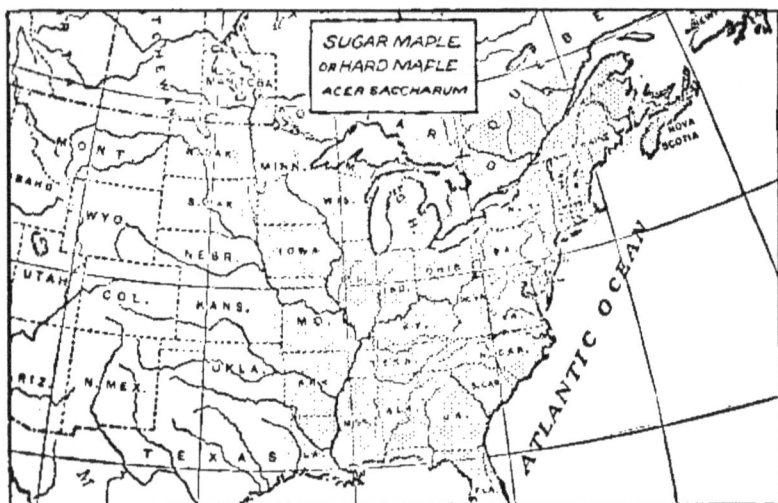

SUGAR MAPLE, ROCK MAPLE OR HARD MAPLE. (*Acer saccharum*)

A large, splendid forest tree, 80 to 120 feet high; red in autumn. Wood hard, strong, tough and heavy but not durable. A cubic foot weighs 43 lbs. It enjoys with Beech, Hickory, etc., the sad distinction of being a perfect firewood. Thanks to this it has been exterminated in some regions.

Bird's-eye and curled Maple are freaks of the grain. Leaves 3 to 5 inches long. Its sap produces the famous maple sugar. This is the emblem of Canada.

There is a black barked variety called Black Sugar Maple (*A. nigrum*). It is of doubtful status.

SILVER MAPLE, WHITE OR SOFT MAPLE. (*Acer saccharinum*)

Usually a little smaller than the Sugar Maple and much inferior as timber. Wood hard, close-grained. A cubic foot weighs 33 lbs. Leaves 5 to 7 inches long. This tree produces a little sugar. It is noted for its yellow foliage in autumn.

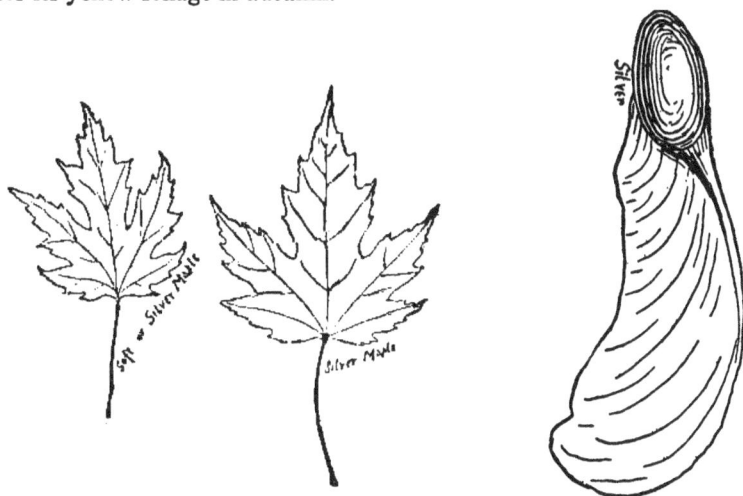

RED, SCARLET, WATER OR SWAMP MAPLE. (*Acer rubrum*)

A fine tree the same size as the preceding. Noted for its flaming crimson foliage in fall, as well as its red leafstalks, flowers and fruit earlier. Its wood is light-colored, tinged reddish, close-grained, smooth with varieties of grain, as in Sugar Maple; heavy. A cubic foot weighs 39 lbs. Leaves 2 to 6 inches long. Produces a little sugar. In the woods there is a common bush 3 to 6 feet high, with leaves much like those of this maple, but the bush has berries on it, it is called the Maple-leaved Viburnum (see later).

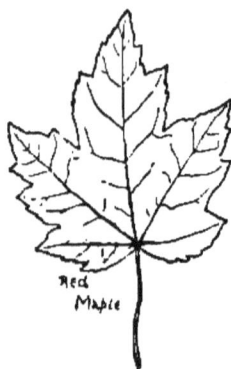

"A small Red Maple has grown, perchance, far away at the head of some retired valley, a mile from any road, unobserved. It has faithfully discharged all the duties of a maple there, all winter and summer neglected none of its economies, but added to its stature in the virtue which belongs to a maple, by a steady growth for so many months, and is much nearer heaven than it was in the spring. It has faithfully husbanded its sap, and afforded a shelter to the wandering bird, has long since ripened its seeds and committed them to the winds. It deserves well of mapledom. Its leaves have been asking it from time to time in a whisper, 'When shall we redden?' and now in this month of September, this month of traveling, when men are hastening to the seaside, or the mountains, or the lakes, this modest maple, still without budging an inch, travels in its reputation — runs up its scarlet flag on that hillside, which shows that it finished its summer's work before all other trees, and withdrawn from the contest. At the eleventh hour of the year, the tree which no scrutiny could have detected here when it was most industrious is thus, by the tint of its maturity, by its very blushes, revealed at last to the careless and distant traveler, and leads his thoughts away from the dusty road into those brave solitudes which it inhabits; it flashes out conspicuous with all the virtue and beauty of a maple — *Acer rubrum*. We may read its title, or rubric, clear. Its virtues not its sins are as scarlet." (*Thoreau*.)

"Never was a tree more appropriately named than the Red Maple. Its first blossom flushes red in the April sunlight, its keys ripen scarlet in early May, all summer long its leaves swing on crimson or scarlet stems, its young twigs flame in the same colors and later, amid all the brilliancy of the autumnal forest, it stands preëminent and unapproachable." (*Keeler*.)

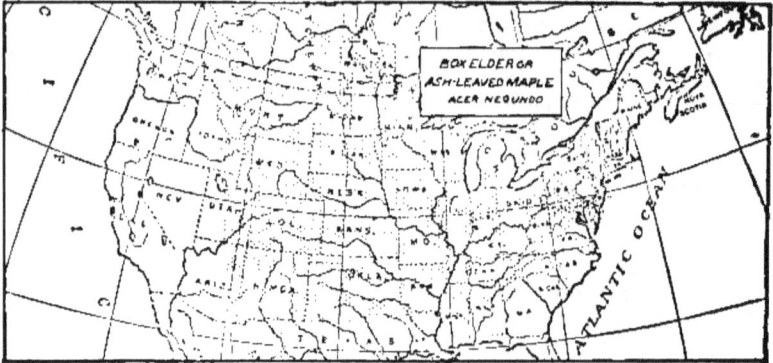

BOX ELDER OR ASH-LEAVED MAPLE. (*Acer Negundo*)

A small tree, 40 or 50 up to 70 feet high, found chiefly along streams. Wood pale, soft, close-grained, light. A cubic foot weighs 27 lbs. Poor fuel. Makes paper-pulp. Leaflets 2 to 4 inches long. Sap yields a delicate white sugar. Chiefly in Mississippi Valley and north to Manitoba, but in the eastern states as an escape from cultivation.

"It was usual to make sugar from maples, but several other trees were also tapped by the Indians. From the birch and ash was made a dark-colored sugar, with a somewhat bitter taste, which was used for medicinal purposes. The box-elder yielded a beautiful white sugar, whose only fault was that there was never enough of it." ("Indian Boyhood," p. 32, by Charles A. Eastman.)

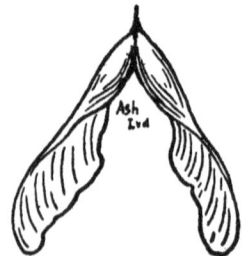

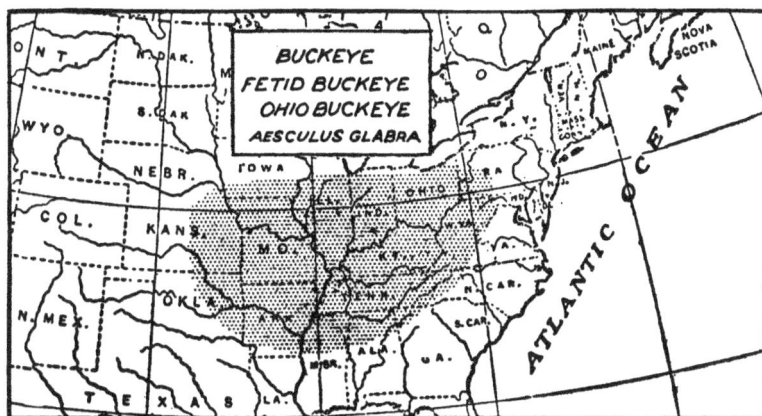

19. ÆSCULACEÆ — BUCKEYE FAMILY

BUCKEYE, FETID BUCKEYE, OHIO BUCKEYE. (*Æsculus glabra*)

Not a large tree, up to 50 feet high. So called because the dark brown nut peeping from the prickly husk is like the half-opened eye of a buck. Leaflets 5, rarely 7, 3 to 6 inches long. Wood, soft, close-grained, light. A cubic foot weighs 28 lbs. Sapwood darkest, used for wooden legs and dishes.

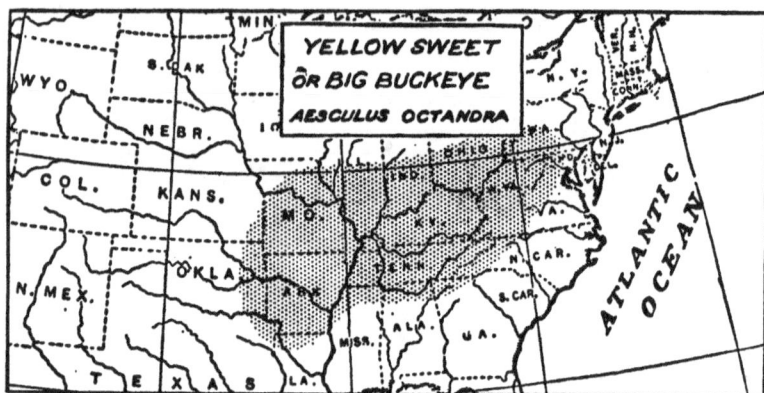

YELLOW SWEET OR BIG BUCKEYE. (*Æsculus octandra*)

A good-sized tree; up to 90 feet high. "Sweet" because its bark is less ill smelling than that of its kin. (*Keeler.*) Wood, soft and white, 27 lbs., per cubic foot, husk of nut, smooth — leaflets 5, rarely 7, 4 inches long; 2 to 3 inches wide.

[HORSE CHESTNUT OR BONGAY. (*Æsculus Hippocastanum*)

A large tree sometimes 100 feet high. Wood, soft, white, close-grained; poor timber. Leaflets 5 to 7 inches long. A foreigner; now widely introduced in parks and roadsides; named either as "horse-radish," "horse-fiddle" and "horse bean" were through using the word "horse" to mean large and coarse, or possibly because the scars on the twigs look like the print of a horse's hoof.

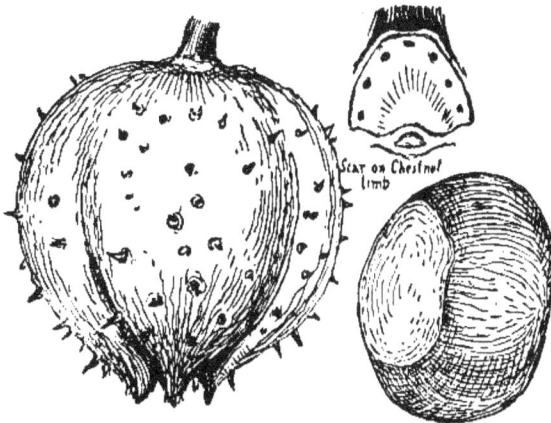

Scar on Chestnut limb

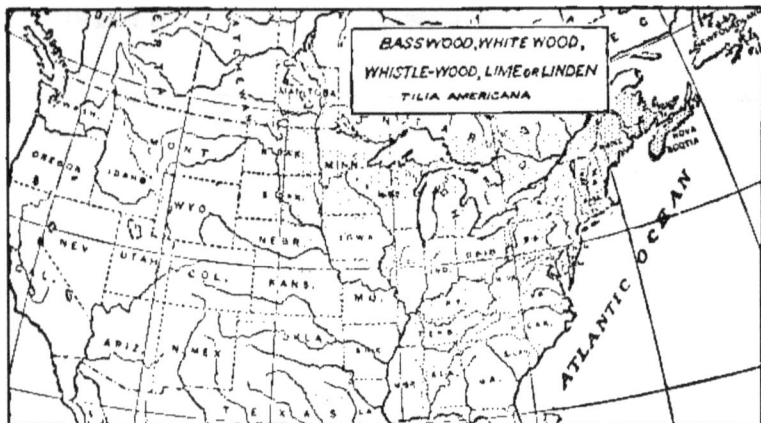

BASSWOOD, WHITE WOOD,
WHISTLE-WOOD, LIME OR LINDEN
TILIA AMERICANA

20. TILIACEÆ — LINDEN FAMILY

BASSWOOD, WHITE-WOOD, WHISTLE-WOOD, LIME OR LINDEN. (*Tilia americana*)

A tall forest tree 60 to 125 feet high; usually hollow when old. Wood soft, straight-grained, weak, white, very light. A cubic foot weighs 28 lbs. It makes a good dugout canoe or sap trough. The hollow trunk, split in halves, was often used for roofing (see log-cabin). Poor firewood, and soon rots; makes good rubbing sticks for friction fire. Its inner bark supplies coarse cordage and matting. Its buds are often eaten as emergency food. Leaves 2 to 5 inches wide. Its nuts are delicious food, but small.

There are two other species of the family, Southern Basswood (*T. pubescens*) known by its small leaves and the Bee tree (*T. heterophylla*) known by its very large leaves.

Basswood Whistle. Take a piece of a young shoot of basswood, smooth and straight, about 6 inches long, without knots, cut it as shown in Fig. 1. Hammer this all around with a flat stick or roll it between two flat boards. Very soon the bark can be slipped off in one whole piece. Now cut the stick to the shape of a whistle plug, slip the bark on again and you have a whistle.

Make it longer and cut off the plug, add holes and you have a pipe.

The exquisite spotless purity of the wood laid bare when the bark is slipped off is so delicate and complete that a mere finger touch is a defilement. It is from this we get the phrase "clean as a whistle."

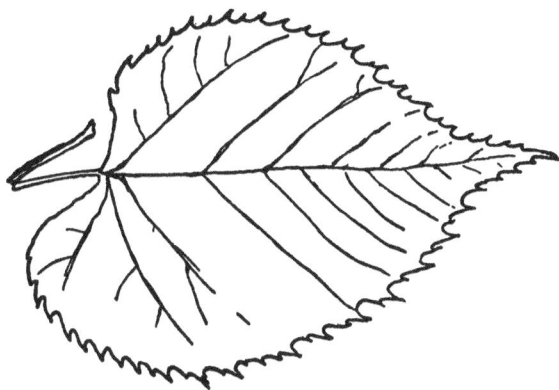

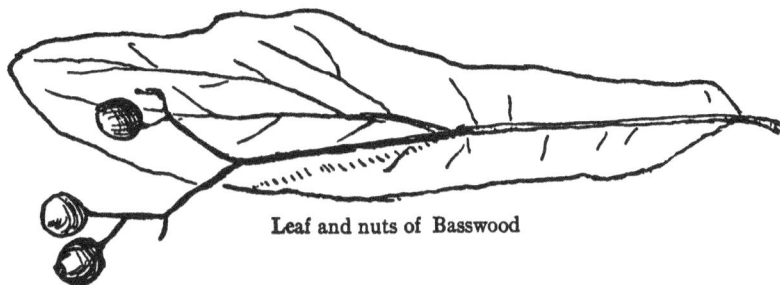

Leaf and nuts of Basswood

Nut, life size

FLOWERING DOGWOOD, ARROW-WOOD, BOXWOOD, CORNELIAN TREE
CYNOXYLON FLORIDUM

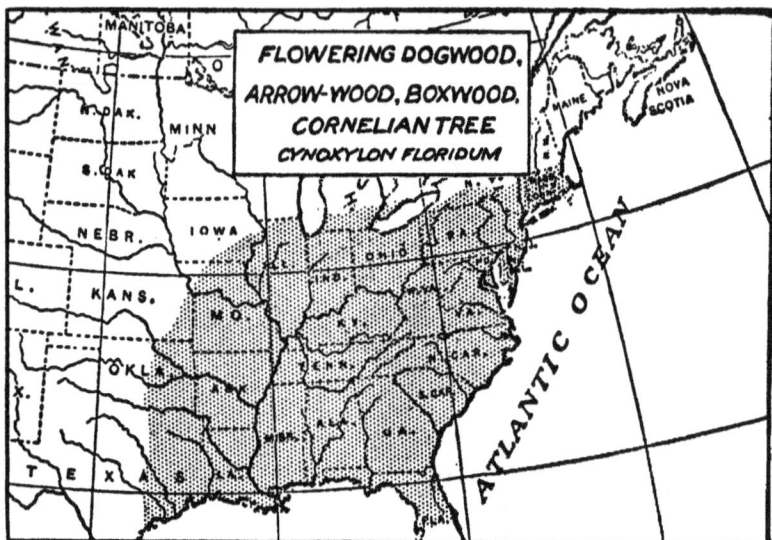

21. CORNACEÆ — DOGWOOD FAMILY

FLOWERING DOGWOOD, ARROW-WOOD, BOXWOOD, CORNELIAN TREE.
(Cynoxylon floridum)

A small tree 15 to 20 feet, rarely 40, with bark beautifully pebbled or of alligator pattern. Wood hard, close, tough, strong, and heavy, a cubic foot weighing 51 lbs. Noted for its masses of beautiful white bloom in spring. A tea of its roots is a good substitute for quinine. Leaves 3 to 5 inches long.

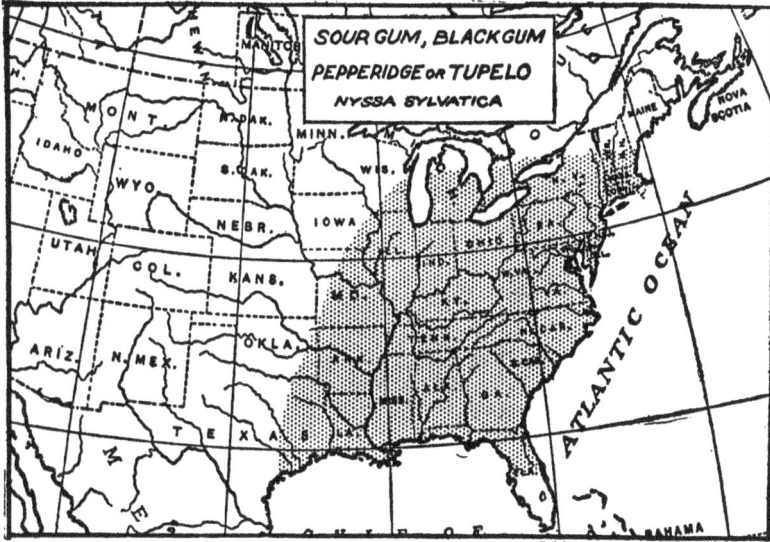

SOUR GUM, BLACK GUM, PEPPERIDGE OR TUPELO. (*Nyssa sylvatica*)

A forest tree up to 110 feet high; in wet lands. Wood pale, very strong, tough, unsplittable and heavy. A cubic foot weighs 40 lbs. Used for turner work, but soon rots next the ground. Leaves 2 to 5 inches long. Noted for its brilliant fiery autumn foliage.

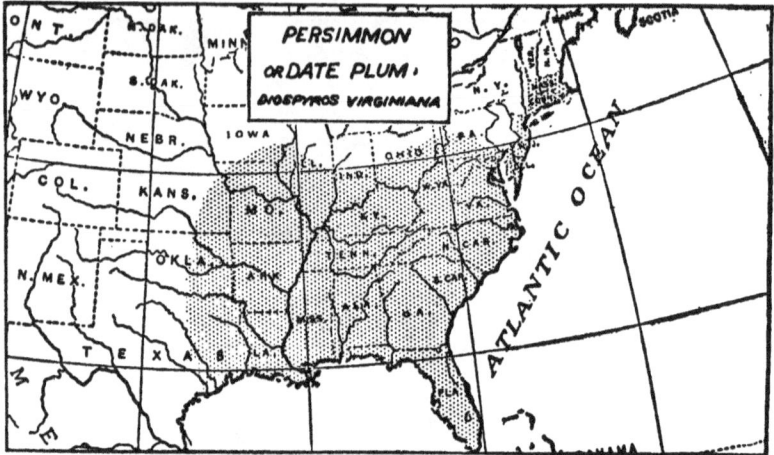

22. EBENACEÆ — EBONY FAMILY

PERSIMMON OR DATE PLUM. (*Diospyros virginiana*)

A small tree 30 to 50 feet high, famous for the fruit so astringent and puckery when unripe, so luscious when frosted and properly mature. Leaves 4 to 6 inches long.

"In respect to the power of making heartwood, the Locust and the Persimmon stand at the extreme opposite ends of the list. The Locust changes its sapwood into heartwood almost at once, while the Persimmon rarely develops any heartwood until it is nearly one hundred years old. This heartwood is extremely close-grained and almost black.

Really, it is ebony, but our climate is not favorable to its production."
(*Keeler*.) Wood very heavy, dark and strong, a cubic foot weighs
49 lbs. Rhode Island to Florida and west to Ohio and Oklahoma where
it becomes a tall tree.

23. OLEACEÆ, OLIVE FAMILY (INCLUDING THE ASHES)

White Ash. (*Fraxinus americana*)

A fine forest tree on moist soil: 70 to 90 or even 130 feet high.
Wood pale brown, tough, and elastic. Used for handles, springs, bows,
also arrows and spears; heavy. A cubic foot weighs 41 lbs. Soon rots
next the ground. Yellow in autumn; its leaflets have stalks, noted for
being last to leaf and first to shed in the forest. Called white for the
silvery undersides of the leaves; these are 8 to 12 inches long; each leaflet
3 to 5 inches long.

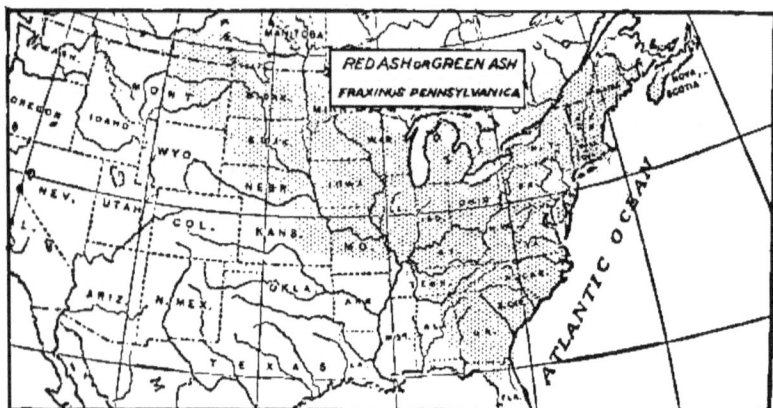

RED ASH OR GREEN ASH.   (*Fraxinus pennsylvanica*)

A small tree rarely 80 feet high. Wood light brown, coarse, hard, strong, brittle heavy. A cubic foot weighs 44 lbs. The Red Ash is downy on branchlet, leaf and leaf-stalk while the White Ash is in the main smooth, otherwise their leaves are much alike. The Green is a variety of the Red.

Leaf and seeds of Red Ash

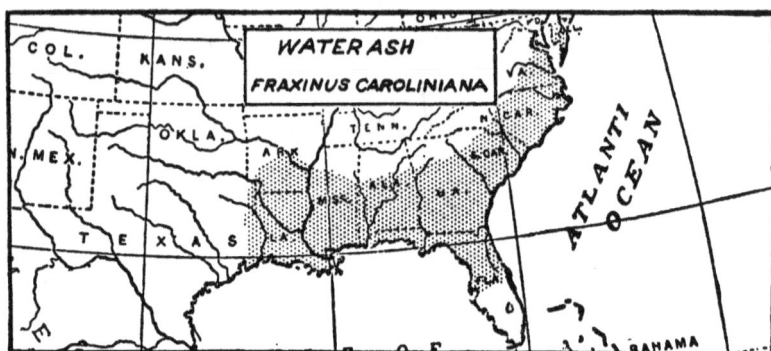

WATER ASH. (*Fraxinus caroliniana*)

A small tree rarely over 40 feet high. Wood whitish soft, weak. A cubic foot weighs 22 lbs; leaflets 5 to 7, or rarely 9; 2 to 5 inches long. In swamps and along streams.

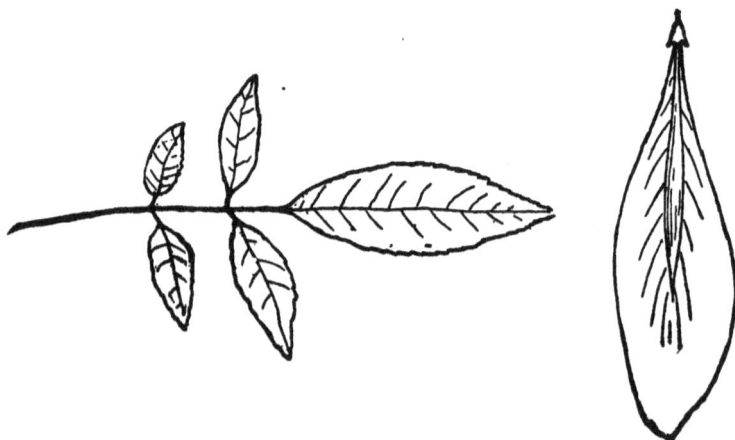

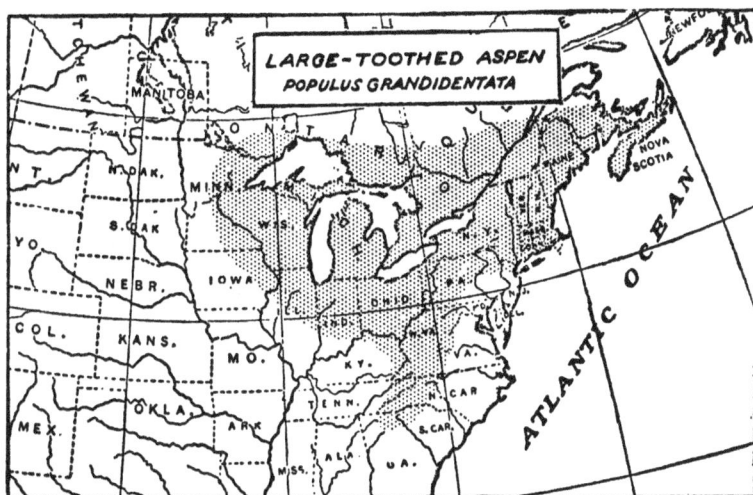

LARGE-TOOTHED ASPEN. (*Populus grandidentata*)

A forest tree, occasionally 75 feet high. Bark darker and rougher than preceding; readily distinguished by saw-toothed leaves. Wood much the same, but weighs 29 lbs. Leaves 3 to 4 inches long.

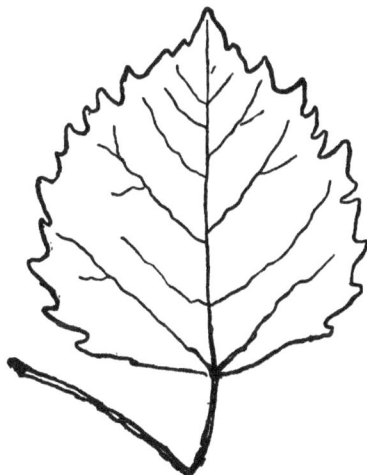

BLACK ASH, HOOP ASH OR WATER ASH. (*Fraxinus nigra*)

A tall forest tree of swampy places; 70, 80 or rarely 100 feet high. Wood dark brown, tough, soft, course, heavy. A cubic foot weighs 39 lbs. Soon rots next to the ground. Late in the spring to leaf, and early to shed in the fall. The leaves are 12 to 16 inches long; its leaflets except the last have no stalk, they number 7 to 11, are 2 to 6 inches long.

Sometimes called Elder-leaved Ash because its leaves somewhat re-
semble the leaves of the Elder, but they are much larger and the leaflets
of the latter have slight stalks, especially those near the base and are
on a succulent green stem which is deeply grooved on top. The thick
bumpy twigs of the Black Ash with the black triangular winter buds
are strong characters at all seasons.

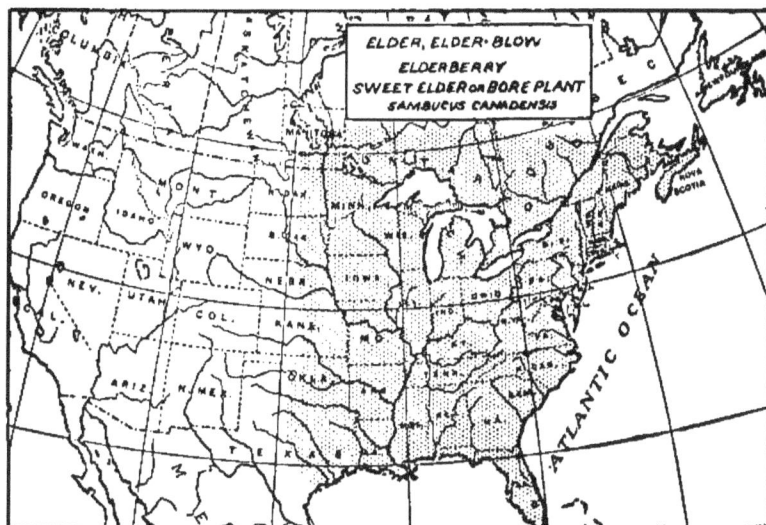

24. CAPRIFOLIACIÆ— HONEYSUCKLE FAMILY

ELDER, ELDER-BLOW, ELDERBERRY, SWEET ELDER OR BORE PLANT.
(*Sambucus canadensis*)|

A bush 4 to 10 feet high, well known for its large pith which can be
pushed out so as to make a natural pipe, commonly used for whistles,

squirts, etc. Its black sweet berries are used for making wine. Its leaves
are somewhat like those of Black Ash, but have a green succulent
stalk. A tea of the inner bark is a powerful diuretic. The young leaf-
buds are a drastic purgative; they may be ground up and taken as
decoction in very small doses. The leaves are 8 to 12 inches long; leaflets,
5 to 11, usually 7, and 2 to 5 inches long. There is another species with
red berries. It is called the Mountain Elder (*S. pubens*) and is found
from New Brunswick to British Columbia, and southeast to California
and Georgia. It has orange pith and purple leafstalks whereas *Canadensis* has yellow pith and green leafstalks.

RED-BERRIED
ELDER
SAMBUCUS RACEMOSA

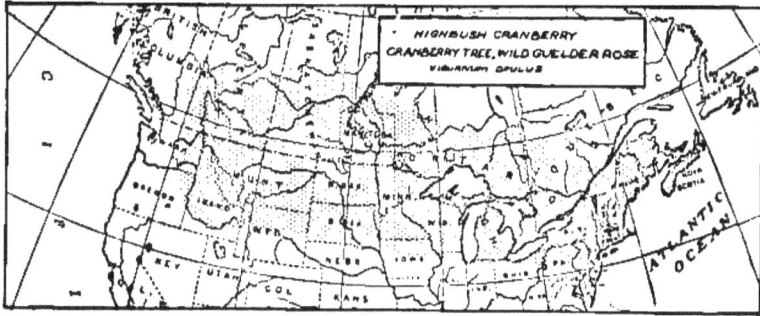

HIGH BUSH CRANBERRY, CRANBERRY TREE, WILD GUELDER ROSE.
(Viburnum Opulus)

A bush 10 to 12 feet high. Noted for its delicious acid fruit, bright red transculent and in large bunches, each with a large flat seed. Leaves 2 to 3 inches long. Found in low grounds from New Brunswick to British Columbia. South to New Jersey, also in the Old World.

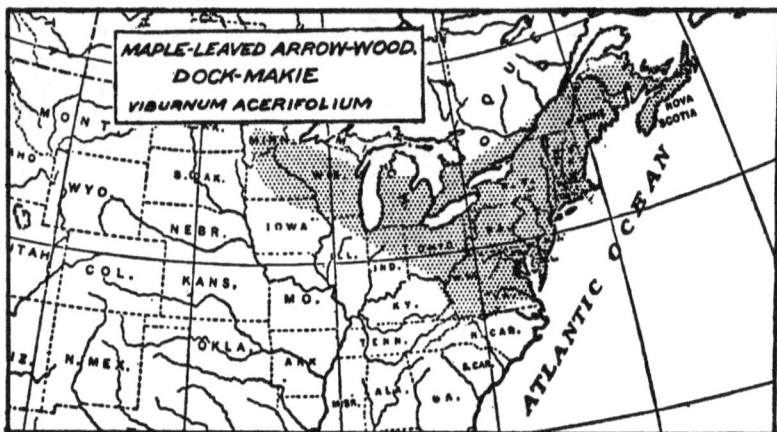

MAPLE-LEAVED ARROW-WOOD, DOCK-MAKIE.   (*Viburnum acerifolium*)

A forest bush, 3 to 6 feet high.   Chiefly noted because of its abundance in the hard woods where it is commonly taken for a young maple.   The style of its leaves however distinguish it, also its berries, these are black with a large lentil-shaped seed.   Leaves 3 to 5 inches long.

ARROW-WOOD.   (*Viburnum dentatum*)

A forest bush, up to 15 feet high; its remarkably straight shoots supplied shafts for the Indian's arrows.   Leaves 2 to 3 inches long.   Its berries blue-black, with a large stone grooved on one side and rounded on the other.   In moist soil.

NANNY-BERRY, NANNY-BUSH
SHEEP-BERRY, BLACKTHORN
SWEET VIBURNUM
VIBURNUM LENTAGO

NANNY-BERRY, NANNY-BUSH, SHEEP-BERRY, BLACKTHORN, SWEET VIBURNUM. (*Viburnum Lentago*)

A small tree, up to 30 feet high. Noted for its clusters of sweet rich purplish-black berries, each half an inch long, but containing a large oval, flattened seed. Leaves 2 to 4 inches long. Wood hard, a cubic foot weighs 45 lbs. It is the largest of the group.

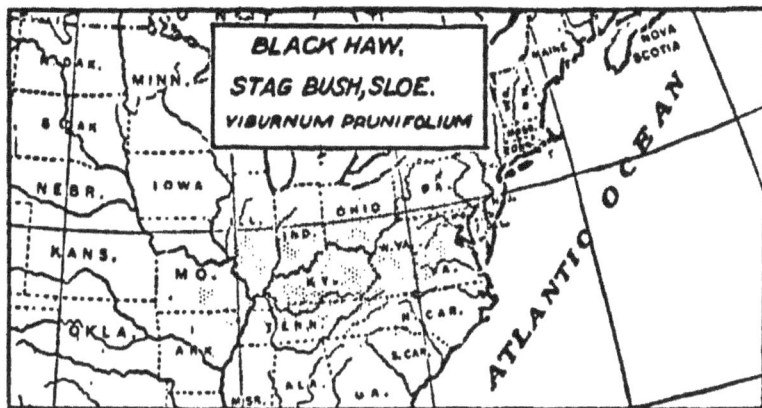

BLACK HAW, STAG BUSH, SLOE. (*Viburnum prunifolium*)

A small tree up to 20 or 30 feet high, much like the Nanny-berry; fruit black, sweet and edible.   Leaves 1 to 3 inches long.   Wood hard, a cubic foot weighs 52 lbs.   In dry soil.

THE END

www.ingramcontent.com/pod-product-compliance
Lightning Source LLC
Chambersburg PA
CBHW020251290326
41930CB00039B/713